Division by Zero

Natural Math

PETER BOCZAR

Creativity and the Conundrum of Counting

Division by Zero

Natural Math

ISBN
978-988-75158-0-7 (Print)
978-988-75158-1-4 (eBook)

Distributed by

MobiShow Ltd.

GPO 4616

HONG KONG

peterboczar@yahoo.com

Other Books by Peter Boczar

Lotus Eater: An Eric Ketch Thriller

Crescent Shadow: An Eric Ketch Thriller

Face Off Nicaragua: An Eric Ketch Thriller

Searching for Uncle Larry: A Missing Airplane Mystery

For
Reverend Paul Manning
Theologian, Mathematician, Inspiration

About the Author

Peter Boczar started writing in the 1970s for magazines and newspapers in Hong Kong, where he also dubbed Chinese kung fu movies into English and made a cameo appearance in the Bruce Lee movie *Game of Death*. As a writer, he traveled extensively throughout the world and notably covered conflict zones in the Middle East, Southeast Asia and Central America.

"As far as the laws of mathematics refer to reality, they are not certain; and as far as they are certain, they do not refer to reality"
—Albert Einstein

"God made the integers. All the rest is the work of man"
—Leopold Kronecker

Table of Contents

Preface ix
Introduction 1
Chapter 0: Back to Basics 8
Chapter 1: Knowing One's Place 16
Chapter 2: What's in a Name? 19
Chapter 3: Addition 23
Chapter 4: Subtraction 25
Chapter 5: Multiplication 29
Chapter 6: Division 30
Chapter 7: Zero Tolerance 32
Chapter 8: Division by Zero 35
Chapter 9: Multiplication Revisited 38
Chapter 10: Natural Numbers 41
Chapter 11: Integers 42
Chapter 12: Rational Numbers 43
Chapter 13: Irrational Numbers 46
Chapter 14: Pi 48
Chapter 15: Euler's Constant 53
Chapter 16: Real Numbers 59
Chapter 17: Imaginary Numbers 60
Chapter 18: Reality Check 61
Chapter 19: Fibonacci Sequence 62
Chapter 20: The Golden Ratio 65
Chapter 21: Spirals 76
Chapter 22: Sets 81
Chapter 23: Infinity & Beyond 83
Chapter 24: Symbols 88
Implications: Natural Math 92
Appendix A: Out-of-the-Box 98
Appendix B: Golden Rectangles 99
Appendix C: Number Operator Properties 100
Appendix D: Division by Zero Table 105

Preface

Mathematicians come in two flavors.

"Pure" and "Applied".

Pure mathematicians study the properties of numbers and their relationships for the sake of itself, while applied mathematicians use math to solve real-world problems.

I often thought that pure mathematicians were just self-indulgent, intellectual snobs, playing a sophisticated game, not bothered about whether their work had any benefit, relevance or application to the real world.

However, over the years, I've come to appreciate the role of pure mathematicians. They do the heavy lifting of math by solving or expanding the intricacies of a very complex system so that others may apply it to their needs.

It is the same relationship that theoretical physicists have with applied physicists or engineers.

Hopefully, *Division by Zero* is not just a bridge between theory and practice, but between the conceptual and natural world.

Not just in math, but how we think about our daily lives and the principles that govern them.

Introduction

This is not really a book about math or the history of math.

I am neither a mathematician nor a historian. I've studied mathematics up through university and read history.

That's about it.

It's a book about thinking creatively.

Math is just the basis for the discussion.

Some people claim you are either born creative or you are not. I don't accept that. *Everyone* is inherently creative.

It's just a question of applying it.

Creativity means leaving your bubble, your comfort zone, your routine and trying to do something differently. It can be something as simple as finding a new way to walk home or opening a soda bottle.

It's about thinking beyond pre-conceived, accepted assumptions, behaviors and boundaries.

But you really need to *want* to be creative.

It's just too easy and lazy to accept established norms or routines and do as you're told. Afterall, if you do or think something different, then you have to accept responsibility for it as well as its consequences. If you just follow along, you can blame the norms, the system or someone else.

I was born with an insatiable curiosity and inquisitiveness. My earliest memories go back to age two. I typically took apart my toys to try to figure out how they worked.

I carried that inborn inquisitiveness throughout my life.

Existence is not enough.

I want to know.

I see life as an ongoing opportunity to explore, discover, learn and grow.

I don't know why. It just feels good.

And as we know by observing nature, when things stop growing, they die.

I was a fairly obedient kid through most of my life. I say fairly, because obedience was sometimes at conflict with my need to know. I could accept the strictest of discipline if there was an explanation for it.

If there was no reason given, or none I could find, then I was troubled.

Also, I sometimes found that the "person-in-charge" was not satisfied if you just followed their rules. They also wanted you to validate their belief system.

Obedience was not enough.

They wanted you to believe.

Creative thinking is often called "out-of-the-box" thinking.

Some of us have been challenged to think "out-of-the-box" by the party puzzle of the same name. That is, connect nine dots arranged in a box-like matrix, using only four lines and without taking your pen off the paper. We typically get hung up trying to find a solution that fits within the perceived box and outlines the boundary of the dots.

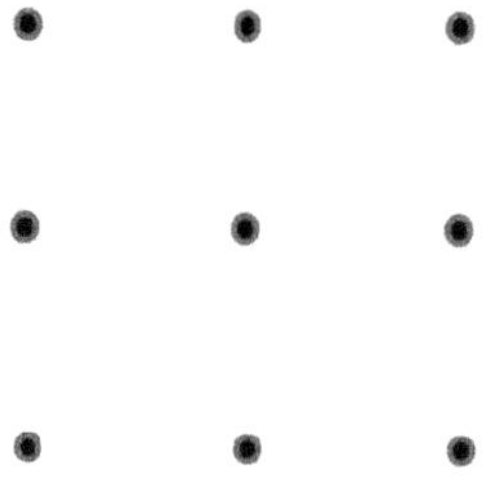

• • •

• • •

• • •

Quite frankly, it's not that easy even if you do think outside the box.

But it does make the point.

Literally.

Solution in *Appendix A*.

Another challenge is to stand three people at the end of a carpet and put an apple in the middle. Then task them to secure the apple using only their bodies and mutual cooperation.

Very amusing at cocktail parties.

Give them a few minutes contorting themselves, trying to stretch their reach by climbing on top of each other and performing other acrobatics. Then, if they don't see the solution, advise them to just roll up the carpet in front of them and grab the apple.

So, why do I use mathematics to promote creative thinking?

First, math is perhaps one of the most defined and disciplined subjects out there. So, a good target.

Also, it's about an idea I had in elementary school that still bugs me.

When learning math, we first learned definitions, axioms, operations and theorems.

Axioms were defined as "self-evident" truths or beliefs. They were supposed to be obvious and unquestionable, even though they couldn't be proven.

However, once you accepted such definitions and axioms, you got dragged into a complex system of rules and operations based on "logical thinking" which they used to prove its theorems and expand it.

We were encouraged to debate those principles and relationships but only within the framework of the system's definitions and boundaries.

Thinking outside the boundaries of that system got you accused of being illogical, crazy or stupid and ostracized from further discussion.

Division by Zero challenges that sort of thinking.

In everything. Not just math.

Here's some examples.

We typically take note of technology companies for their innovation and futuristic thinking.

However, I've had some different experiences.

I once worked with a telecom company in the days when mobile phones were thick as a brick, clumsy to carry and very expensive to use.

As a result, consumers used them in conjunction with pagers. Users kept the phones in a briefcase while they wore the pagers on their belts.

They texted messages to each other on the pager and connected on the phone only if it was important enough to spend the money on airtime. Airtime was not cheap in those days.

In that regard, I suggested to the company that they should incorporate a text messaging feature into the mobile phones themselves.

They reflexively responded that I didn't understand their business.

After the internet was established, I suggested to another big, international mobile phone company that they should put a modem in their phones that allowed users to connect to the net and get their email.

Their marketing director, who was armed with an expensive consumer research study, brushed me off.

"We've already looked at that," he explained. "Based on our study, consumers think of the internet as a 'lean forward' experience, whereas phones are a 'lean back' experience. You don't understand our business."

The client is always right, so I didn't bother to ask why someone couldn't lean forward when getting email from their phone.

At the same time, I must admit, I never thought of putting a camera in a phone even though I was a photography enthusiast. I didn't see the benefit, nor predict the quality that today's phone cameras might achieve. I also didn't see the big picture. It's not just about taking a photo but capturing and sharing a memory, recording data, purchasing groceries, validating an ID, downloading information, photocopying documents, translating a foreign language and more.

A bit more than taking pictures.

Seeing the big picture is often illustrated by the story of two guys whose boat is washed ashore on a tropical island after a storm with a cargo of shoes.

The first guy goes exploring and dejectedly reports back that the natives run around barefoot. There is no market for their cargo, so they have nothing to trade for food and water.They are doomed.

The second guy sees the opportunity to be the monopoly shoe provider. He goes ashore and convinces the natives that wearing shoes has great benefits. Not just in terms of protecting their feet when walking, but also when running after their prey while hunting.

Shoes mean safer soles and fuller bellies.

The natives buy it.

But then the first guy becomes dejected again after the natives each buy a pair of shoes and see no need for more. It could be years before they wear out and need to be replaced.

Again, starvation looms in his mind.

The second guy goes back and demonstrates how different shoes can be used for different occasions, to enhance one's dress and make life more fun and interesting.

So, what does this have to do with *Division by Zero* and *Natural Math*?

In short, it's about challenging assumptions, seeing the big picture and thinking creatively.

As a young schoolboy, I typically didn't take anything at face value.

I needed to understand all sides of a point of view before I could accept it.

Not for the sake of argument or debate, but for mutual understanding through discussion and dialogue.

I also liked puzzles, riddles and brain teasers. The solutions were typically not the obvious ones.

In elementary school, mathematics was initially straightforward. First came addition, then subtraction, then multiplication.

And then came division.

Something was not quite right about division. It was not that easy to compute and often resulted in remainders or fractions.

Most disturbing was that division by zero was an anomaly.

Some claimed it equaled infinity. Others that it was "undefined", "not allowed" or didn't have a solution. Just leave it alone.

However, I thought everything in mathematics was supposed to be defined and provide a solution. That's what made math, math.

So, this never sat right with me.

More importantly, neither explanation seemed consistent with the natural order of things. And math was supposed to describe nature.

Right?

Eventually, I came to my own conclusions, presented here.

This is not just a book about zero or math, but about discovery, thinking outside the box, beyond the rules and challenging oneself to find and consider innovative alternatives.

It is about thinking creatively about any subject, even one that sometimes seems strictly fenced in by rules such as math.

I start with some historical background for context as well as fun. However, I do not spend a lot of time delving into the history of numbers or mathematical principles. I am not a historian and my research into the development of numbers and mathematics found that it is not always consistent.

Rather, I'm looking to share an idea, not investigate the past. Hence it should be a quick read. Feel free to skip ahead if something hangs you up. Especially some of the math.

The first seven chapters set up the idea of *Division by Zero* and chapter eight explains it. So, don't skip this one. A discussion of some key mathematical definitions and concepts follows. It looks to evaluate those concepts in terms of their relationship to the natural world. Notably the "Big Three".

Pi, *Phi* and *e*.

Subsequently, I raise the issue of infinity, which is as problematic as zero, address the importance of symbols and then bring everything to closure.

There's also lots of diagrams to help you through.

In elementary school, we put a slash through the zero numeral to clarify that our written symbol represented the number and not the letter "O". Especially important for those of us with sloppy handwriting.

Hence, I introduce the following symbol to represent the numerical operation of *division by zero*:

The graphic not only clarifies the number from a letter but speaks to the idea of division. So, I thought it appropriate to use as the symbol for my concept.

I pronounce the operation "zed".

"Zed" is also how radio operators pronounce the letter "Z" to distinguish it from "C" when spelling out a word. Some English speakers just pronounce it that way in their common practice. And afterall, *zed* does represent the first letter of "zero".

So, just like we'd say "three *times* four" for (3 x 4), *I'd* say "three *zed* four" for the operation (3 Ø 4).

Chapter 0: Back to Basics

In the beginning there was no zero.

There was no need.

Counting started with one and the act of counting was just ordering items based on their quantities.

One rock, one dinosaur, one seashell. One, two, three, four, five fingers. Ten including both hands or twenty including the toes on both feet.

You didn't count zero rocks. If there weren't any rocks, there was no need to count them because there were simply none.

Early man likely counted by using rocks, shells or sticks.

And he didn't need a counting system or written symbols to compare quantities.

He only needed to line up his pebbles one-to-one and see which side contained more items. For example, matching one group of dots against another shows that Group B is bigger than Group A. No need to count. It's easy to see:

A: ● ● ●

B: ● ● ● ● ●

Later, early man likely used tally marks to represent numbers. The tally marks were probably scratched in the dirt, then on sticks or bones.

Somewhere along the line, man found it convenient to assemble the tally marks into groups. Perhaps groups of five. This seemed natural. Afterall, a group of five fingers made a hand.

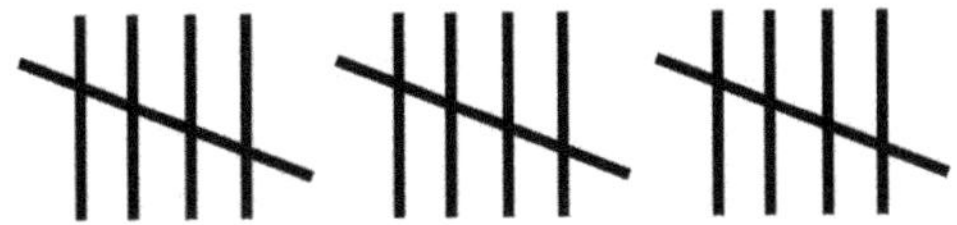

Crossing off four lines with a diagonal stroke into groups of five made it visually easier to count.

If you counted all your fingers and toes, you'd get 20. We could also take the lines in groups of four.

You might get this if you count the spaces between fingers, instead of the fingers themselves. If you include spaces between the toes and fingers you get 16.

In fact, you can express any total regardless of how you group the units. Take the number eleven.

Groups of five give you:

Groups of four give you:

Mathematicians call such groupings a *Base*. That is, a counting system that groups its digits and associates them with a specific value.

We unknowingly learned about *Bases* when teachers had us counting by 2s, 3s, 4s, 5s...

The instruction to "count by twos" meant reciting "two, four, six, eight..."

After a bit of practice, we didn't even have to think about it. The numbers in the series came out automatically.

After counting by 2s, the easiest for me was counting by 5s. Maybe because every number ended in zero or five. "Five, ten, fifteen, twenty, twenty-five, thirty..."

Base 5 was an obvious candidate for our mainstream number system. Man had 5 fingers on each hand and 5 toes on a foot.

Base 4 might have been another candidate.

Base 4?

As already noted, counting the spaces between the fingers gives you the number 4.

But another way to get 4 is to count your fingers by tapping your thumb to your fingertips.

Try it. You can do it using only one hand.

Then somewhere along the way *Base 12* popped up, also called the *duodecimal* system.

This gave us 12 inches in a foot, 12 months of the year, 12 hours on a clock face, 12 tribes of Israel, 12 Apostles, 12 zodiac signs, 12 Knights of the Round Table, 12 days of Christmas, 12 of anything in a dozen. Notably a dozen eggs and a dozen oysters.

Where did we get that? Maybe it was another way of counting on your hands with your thumbs.

Instead of just tapping your thumb to your fingertips, tap it to each joint. There are 3 joints in each of your fingers except the thumb which only has two. You'll find you're tapping 12 times on each hand.

And if early man looked up into the night sky, he'd find that there are 12 cycles of the moon.

Maybe it's nature's way of indicating *Base 12* might be the most *natural* way to count when describing the world.

Yet, we didn't divide an inch into 12 units but into successive half units. So: ½, ¼, 1/8, 1/16, 1/32.

Why?

Maybe, it was just easier to cut things in half.

Meanwhile, some of the earliest civilizations, such as the Sumerians and Babylonians, used *Base 60.* Also called the *sexagesimal* counting system.

Where did they get that?

I can't think of anything in nature that appears in groups of 60 units.

If you thumb-count all the joints on your fingers, you get 24. Including the toes, but excluding the big toes, you get 48. If you include the thumb and big toe joints, you get another 8 digits for a total of 56.

So, now what?

Let's think just a little bit differently about how early man might have counted using his fingers.

Separate your hands and open the left one, but close the right one into a fist.

Use the left hand to count the joints using your thumb as explained before.

Start by tapping your left thumb to the bottom joint of your index finger, then count the joints until reaching the last joint on the tip of your little finger. You get 12.

Now extend one finger on the right hand to register one cycle of counting 12 joints and start all over again until you run out of fingers on your right hand.

You will end up counting 5 cycles of 12 units, or 60 in total.

Wow!

It must have made a big impression, as vestiges of the *Base 60* counting system made it to the modern world.

Clocks are the most obvious example. There are 60 seconds in a minute and 60 minutes in an hour.

But wait. Doesn't the clock use a mix of *Bases?*

We divide one hour of 60 minutes into a half-hour of 30 minutes, then half of a half-hour or quarter-hour into 15 minutes. But then something strange happens. Instead of dividing a quarter-hour of 15 minutes in half, we divide it into thirds. So, each numeral on the clock face measures 5 minutes, and there are two rounds of 12 hours in our 24-hour day.

Clock time seems to be a mixture of *Base 60*, *Base 12*, *Base 30*, *Base 15* and *Base 5*.

Interestingly, all these numbers divide evenly into *Base 60*. Along these lines, some have suggested that a *Base 60* counting system allowed ancient civilizations to easily divide up quantities without resorting to fractions. It was the smallest whole number that had the greatest number of divisors. So, using it made the math easier, especially for merchants computing business transactions.

If you exclude the number one and the number itself, 60 has 10 divisors. The nearest competitors only have 7 or 8.

Divisors

	?	2	3	4	5	6	8	9	10	12	15	16	18	20	24	30
	10	X			X											
	12	X	X	X		X										
	15		X		X											
	16	X		X			X									
Base	18	X	X			X		X								
	20	X		X	X				X							
	24	X	X	X		X	X			X						
	36	X	X	X		X		X		X			X			
	48	X	X	X		X	X			X		X			X	
	60	X	X	X	X	X			X	X	X			X		X

Meanwhile, the Egyptians used *Base 10*. It's an obvious choice. We have 10 fingers on both hands and 10 toes on both feet.

Elsewhere, Central American civilizations, such as the Mayans used *Base 20*, also called the *vigesimal* system. This seems natural. Afterall, 10 fingers and toes combined make 20 units.

We even have vestiges of this in modern counting language.

The number eighty in French is *quatre-vingts* or four twenties. And ninety is *quatre-vingt dix* or four twenties plus ten. But there's also a bit of *Base 60* in French as the number seventy is *soixante-dix* or sixty plus ten.

As a student some time ago, I learned of US President Abraham Lincoln's *Gettysburg Address* in 1863 which began, "Four score and seven years ago…" referencing the American *Declaration of Independence* of 1776.

Score designated 20 years, so the math meant 4 groups of 20 or $(20 \times 4) = 80$. And $(80 + 7) = 87$, which equals $(1863 - 1776)$.

The old British monetary system seemed to use a mix of *Bases*.

It included *farthings*, *pence*, *shillings*, *pounds*.

Pence was a bit complicated. It was the plural of *penny* however, only when referring to a monetary amount such as *seven pence*. If referencing the number of coins, the plural of *penny* was *pennies*. For example, there are six *pennies* on the table.

A monetary *penny* was designated by the lowercase, letter "d" referencing the Latin *denarius* coin of ancient Rome. So, eight *pence* was written 8*d*.

It's Greek to me.

Stepping back to the beginning, four *farthings* made one *penny*. Then 12 *pence* made a *shilling* and 20 *shillings* made a *pound*.

So, 240 *pence* and 960 *farthings* made a *pound*.

There were also *two-pence* and *three-pence* coins, respectively called *tuppence* and *thruppence*. And a *four-pence* coin called a *groat*.

Groan!

Why not a *quippence*? Afterall, the British are famous for their witty quips.

There was also a *crown* which was one-quarter of a *pound* and a *half-crown* which was one-eighth of a *pound*.

The largest common divisor of all these units is 4, so maybe the British were thumb-counting their fingers and inadvertently using *Base 4*.

The British went decimal in 1971 and further confused everyone. People carried around calculation tables so they could understand the value of the new money in terms of the old.

However, the British still use a unit of weight called a *stone*. One *stone* is 14 pounds. Someone 180 pounds is noted as weighing 12 *stone*, 12 pounds.

A bit too heavy for me to understand.

The US monetary system also uses mixed *Bases*.

Five pennies make a nickel, ten a dime and one hundred a dollar. Sounds pretty decimal so far, but then five nickels are recognized as a quarter and four quarters make a dollar.

Are four quarters made up of 25 pennies easier to manage than five coins, each with the value of 20 pennies? Then, there are half-dollar coins.

Again, like slicing up inches, maybe it's easier to cut a dollar in half, then in half again instead of trying to cut it into fifths.

Fortunately, the decimal system was accepted internationally for the sciences.

Historically, computers use *Base 2* which consists of 0 and 1 simply because an electronic circuit can be on or off, so represented as 1 or 0.

The first *Base 10* counting numbers are:

0 = 0

1 = 1

2 = 10

3 = 11

4 = 100

5 = 101
6 = 110
7 = 111
8 = 1000
9 = 1001
10 = 1010
And so forth.
So, what about seven?
Seven days in a week.
Seven seas.
Seven continents.
Seven wonders of the world.
Seven colors in a rainbow.
Seven notes in the music scale.
Seven deadly sins.
Enough!
Let's move on.

Ø

Chapter 1: Knowing One's Place

Zero didn't really have a place in counting systems until man found it convenient to arrange his tallies in columns, each representing the *Base* of the underlying counting system.

Ancient man grouped number counters in columns by drawing lines in the dirt or placing pebbles on the ground. Later, he drew the lines or held the stones in a tray of sand. The tray allowed him to take this "calculator" anywhere and the sand kept the marks or counters in place. He later cut grooves into wooden trays to hold the stones, shells or seeds in place.

Later, he put shells, seeds or beads onto rods, which evolved into the *abacus*.

We call this a *place-value* number system because the placement or position of the counter determines its value.

So, the ancients really didn't need a symbol such as zero to separate and define places because they already physically separated the counters by columns in the sand or on the sticks.

They simply left that column empty.

Even a vast civilization like ancient Rome did not see the value of a system using a symbol like zero.

The Roman numerals are sometimes considered a *place-value* system but they are more like a "place-symbol" or "place-numeral" system. As such, we learned them as Roman *numerals* and not Roman *numbers*.

Roman numerals are written left to right, highest to lowest value. However, they have some unique rules.

Initially, one to four were single strokes of *I*. Five was *V*. Ten was *X*. Fifty was *L*. One hundred was *C*. Five hundred *D* and a thousand *M*.

When one value reached the value of the next larger symbol, it transformed into it. So, *IIIII* became *V* and *XXXXX* became *L* or fifty.

However, the system was later simplified to reduce the number of times a symbol repeated to three repetitions. Presumably, to make them easier to read.

To make it work, a rule was introduced such that a smaller number which preceded a larger one was subtracted from it.

So the number four, initially represented as *IIII* was simplified to *IV*. Similarly, *XXXX* or forty became *XL*. Some call these *compound numerals*.

However, even with the simplification, the notation was still pretty cumbersome.

One hundred thirty-four is *CXXXIV* compared to 134 in modern Hindu-Arabic numerals.

Expressing larger numbers is even more complicated.

Nineteen-eighty-four or 1984 is *MCMLXXXIV*.

Separating the numbers by commas or parentheses might have helped. Consider:

M, CM, LXXX, IV
(M) (CM) (LXXX) (IV)

It's easier to translate as one thousand (1,000), nine hundred (1,000 – 100 = 900), eighty (50 + 30), plus four (5 – 1). But that was not the practice.

Simple arithmetic such as addition was also awkward.

For example, think about how you would add *MMCCLXIII* + *MCLXXXI*. That is, 2,263 + 1,181 = 3,444.

One method is to group like symbols under each other, then regroup them and apply the notational rules.

MM	*CC*	*L*	*X*	*III*
M	*C*	*L*	*XXX*	*I*
MMM	*CCC*	*LL*	*XXXX*	*IIII*

This becomes:

MMM CD XL IV

Or,

MMMCDXLIV

Subtraction gets more interesting and don't even think about multiplication.

Yet the Romans built phenomenal architectural structures and commanded an empire covering much of the known world using this cumbersome system.

And the numerals also made it into the modern world.

Clocks and watches often use Roman numerals, notably dress watches. Though they are not consistent in their use of *IIII* or *IV*. London's *Big Ben* uses *IIII*.

Architects also use them to date buildings and publishers use them to mark book chapters. Maybe they are considered more classic and stylish or add a note of authority.

So, how does a *real* place-value system work?

Basically, it's the same for all *Bases*.

Each place represents the value of the *Base* to its next multiple. Let's make it easy and start with *Base 10*. So, for example:

The *numeral* 2,123 in *Base 10* is comprised of 3 units of ones, plus 2 units of 10, plus 1 unit of 100 (10 × 10) and 2 units of 1,000 (10 × 10 × 10).

The same *numeral* in *Base 4* would mean 3 units of ones, plus 2 units of 4, plus 1 unit of 16 (4 × 4), plus 2 units of 64 (4 × 4 × 4). which would have a *value* of 155 in *Base 10*.

Get the idea?

Note that it's very important to recognize the difference between the *numeral* and the *value* it represents in the assigned *Base*. Enough to throw you off base.

Never mind.

We're basically done with places and *Bases*.

Chapter 2: What's in a Name?

The concept of numerals as placeholders was reportedly introduced into the West by the Italian mathematician Leonardo Bonacci, more commonly known as Fibonacci.

Leonardo was born the son of Guglielmo Bonacci in Pisa during the 12th century and his nickname "Fibonacci" allegedly comes from the Latin *filius Bonacci* meaning "son of Bonacci".

However, some claim he was not known by "Fibonacci" in his lifetime. Reportedly, he was just known as "Leonardo of Pisa" or "Leonardo Pisano".

Pisa was one of the Middle Ages, Italian maritime, city-states such as Genoa, Amalfi and Venice that competed with each other for the sea trade throughout the Mediterranean, including that of the Muslim world.

Reportedly, Fibonacci worked for his father as a customs clerk in what is now Algeria where Pisa had an enclave. And he traveled throughout the Muslim world on business where he learned about Arabic numerals and mathematics, including zero.

He is credited with introducing them to Europe through his book *Liber Abaci* or "Book of Calculations".

However, the Muslims most likely had already introduced them in Europe when they occupied the Iberian Peninsula a few hundred years earlier.

Reportedly, the Muslims originally got their numerals along with zero and its *place-value* system from the Hindus, and called them "Hindu numerals". So, we later called them "Hindu-Arabic" numerals to acknowledge their full heritage.

The origin of the name *zero* is not clear.

One popular history is as follows.

The Hindu name for zero was *sunya* meaning "empty" or "nothingness". In Arabic, this was transliterated into *sifr* though it's not clear whether it held the same meaning. Modern Arabic dictionaries just translate this into English as *zero*, *null*, *nil*.

No insight there.

Reportedly, this was "romanized" as *zephyrus* in Latin and *ζεφυρος or zephyros* in classical Greek. However, these translated as "westerly wind" or "light wind" like *zephyr* in English. They did not reference anything related to emptiness or nothingness, so were likely just translated based on pronunciation. Though *Zephyros* was the ancient Greek god of the west wind.

Some literature claims *sifr* was originally Romanized as *zephyrum* in Latin. However, my brief encounter with Latin in junior high school suggests that is the *accusative case* of the noun *zephyrus* which would make it the direct object of a verb. That made no sense to me. So, I stuck with the nominative or noun form, spelled *zephyrus*.

Meanwhile, a modern English language translation of *Liber Abaci* spells it *zephir*, and also claims Fibonacci spelled it *zephirum*. I could not find either word in a Latin dictionary.

Anyway, *zephyrus* and *zephyros* became *zefiro* or *zevero* in various Italian dialects, where it got contracted to *zero*, then adopted in French as *zéro*, which became *zero* in English.

However, like most innovative ideas, it did not get accepted by the mainstream until many years later.

Establishment mathematicians were quite happy with Roman numerals and their abacuses or sand boxes which already separated values by columns. They didn't see the benefit of changing their ways for several centuries.

Also, they claimed that Hindu-Arabic numerals could be easily confused or counterfeited.

The numerals at that time were a bit different from those used today, however similar enough to make the point. Consider the modern numerals we use.

A poorly written "2" might be mistaken for a "3". And a "7" can be reduced to a "4". Then consider how easy it might be to make a "5" into a "6".

Similarly, zero could easily be converted into an "8" quickly increasing the value of a transaction.

In modern times, Fibonacci is best known for his number sequence which he allegedly developed by observing nature. Reportedly, the reproduction of rabbits.

But more on that later.

Meanwhile, zero needed to be figured out.

Was it a *numeral* or a *number*?

As a numeral, it might just be a placeholder such as 10, 105 150, 1050 and so on. And it simply marked the empty space in the sandbox calculating tray.

As a number, well, that is more complicated.

Once you introduce it as a number, you need to define it in the traditional mathematical system and how it operates within the rules.

First of all, is it an *odd* or *even* number?

Some claim it's *even* because $0/2 = 0$ since an *even* number is *defined* as divisible by two. And an *odd* number is *defined* as an *even* number plus one.

So, is it negative or positive?

Good question.

Some mathematicians avoid answering the question by just defining it as "neither". Others allow -0 and $+0$ depending on the situation.

Then, using zero as a number means understanding how to use it when combining or operating on two numbers.

Addition is easy. You add zero or nothing to any number and the original number remains.

For example, $(1 + 0) = 1$.

In general, for any number n, $(n + 0) = n$. Same for subtraction.

So, $(1 - 0) = 1$.

Similarly, for any number n, $(n - 0) = n$.

But things start to get tricky with multiplication. Any number n, multiplied by 0 is *defined* by mathematicians to be 0 or $(n \times 0) = 0$. Does it make sense that *any* number *times* zero should be zero?

Afterall, if zero is nothing, then operating on something by nothing is essentially not operating on it at all. So, it should change nothing.

Let's explore this further by using the simple math tables we learned.

Chapter 3: Addition

Addition is simple. You put a few objects together then count the total. So, if you start with one object,

then add two more,

you get a total of three.

This can be summarized as an addition table.

+	0	1	2	3	4	(5)	6	7	8	9
0	0	1	2	3	4	5	6	7	8	9
1	1	2	3	4	5	6	7	8	9	10
2	2	3	4	5	6	7	8	9	10	11
3	3	4	5	6	7	8	9	10	11	12
(4)	4	5	6	7	8	(9)	10	11	12	13
5	5	6	7	8	9	10	11	12	13	14
6	6	7	8	9	10	11	12	13	14	15
7	7	8	9	10	11	12	13	14	15	16
8	8	9	10	11	12	13	14	15	16	17
9	9	10	11	12	13	14	15	16	17	18

Take the number in the row, move up the column through the addition "operator" noted by the "+" sign, then across to the number you want to "operate" on and find the solution at the intersection of their rows and columns.

Hence, (4 + 5) = 9.

Chapter 4: Subtraction

Subtraction is a little bit more complicated.

If you have three objects,

then subtract or remove two,

only one is left.

No surprises here.

However, subtraction is the first time in the chronology of learning math that we are required to think outside of our natural world.

This time, let's start with one object,

then subtract or "take away" two.

But what's this?

Math class would say you get a *negative* number, specifically −1. However, there's no such thing as a negative object, either in nature or made by man. Show me a negative rock or TV set. Though I have met a number of negative people.

I suppose you could think of this as an item "owed". That is, if you only have one item but someone requests two, then you could say that you "owe" that person one object. Alternatively, you could think of it as an item to be replaced.

My teachers introduced *negative* numbers by using a "number line". This helped visualize the concept. We thought of it simply as counting backwards. So, a *negative* number was not really a number, but a marker.

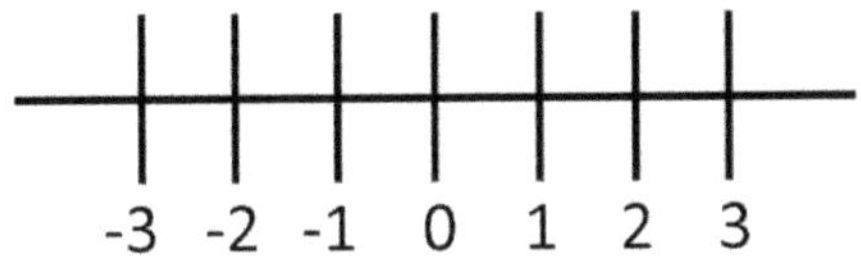

It was still hard for me to visualize within nature and even more difficult to conceptualize in my head. But the number line created a rule, if not a ruler and that made it easier to accept.

At the same time, the number line created another problem.

The calendar.

We typically think of a calendar as being a number line of sorts. But there are no positive or negative years. Only years going backwards and years going forward, arbitrarily designated as *BC* or *AD* and *BCE* or *CE* to be politically correct.

But no zero year.

Why is there no *Year Zero*?

Think of the problem it causes.

Take a person born in 5*BC* and try to calculate how old he is in 3*AD*.

If you assign a negative value to the *BC* years and add arithmetically, you subsequently get (–5 + 3) = –2 years old.

At the same time, if you take the positive *AD* years and subtract the negative *BC* years, you get 3 – (–5) = 8.

Let's try something else.

Between 1*BC* and 5*BC* there are (5 – 1) or four years. Between 1AD and 3AD there are (3 – 1) or two years. This gives his age as (4 + 2) = 6 years.

Poor kid. He'll never know when he reaches drinking age.

-	0	1	2	3	4	5	6	7	8	9
0	0	-1	-2	-3	-4	-5	-6	-7	-8	-9
1	1	0	-1	-2	-3	-4	-5	-6	-7	-8
2	2	1	0	-1	-2	-3	-4	-5	-6	-7
3	3	2	1	0	-1	-2	-3	-4	-5	-6
4	4	3	2	1	0	-1	-2	-3	-4	-5
5	5	4	3	2	1	0	-1	-2	-3	-4
6	6	5	4	3	2	1	0	-1	-2	-3
7	7	6	5	4	3	2	1	0	-1	-2
8	8	7	6	5	4	3	2	1	0	-1
9	9	8	7	6	5	4	3	2	1	0

Chapter 5: Multiplication

Multiplication seems simple at first, but could be tricky if thought of differently.

I learned that multiplication is just a series of additions. So, (2 + 2 + 2) can be expressed as (2 × 3) or (2)(3). They all equal 6. It is addition shorthand. Hence, the famous multiplication table we had to memorize in elementary school.

x	0	1	2	3	4	5	6	7	8	9
0	0	0	0	0	0	0	0	0	0	0
1	0	1	2	3	4	5	6	7	8	9
2	0	2	4	6	8	10	12	14	16	18
3	0	3	6	9	12	15	18	21	24	27
4	0	4	8	12	16	20	24	28	32	36
5	0	5	10	15	20	25	30	35	40	45
6	0	6	12	18	24	30	36	42	48	54
7	0	7	14	21	28	35	42	49	56	63
8	0	8	16	24	32	40	48	56	64	72
9	0	9	18	27	36	45	54	63	72	81

More on multiplication later.

Chapter 6: Division

In elementary school, we learned that division was essentially a form of subtraction. It was the number of times you could remove one group of items from inside another.

So, 6 items divided by 3 meant taking away groups of three from the 6, resulting in 2 new groups, expressed as (6 ÷ 3) = 2 or 6/3 = 2. If the number did not divide evenly, then you got a remainder which could be expressed as a whole number or parts of it.

A fraction.

So, 7/2 = 3 with a remainder of 1, could also be expressed as 3½.

The number you started with was called the *dividend*. The number you divided it by, the *divisor* and the result, the *quotient*. That is (*dividend* ÷ *divisor*) = *quotient*.

Division's association with subtraction became more obvious in the "long hand" method of division, which involved subtracting multiples of the *divisor*.

So, (135 ÷ 4) took you through a process of:

```
    33
  ____
4|135
  12
  ___
   15
   12
  ___
    3
```

Therefore, 135/4 = 33 with a remainder of 3 or $33\frac{3}{4}$.

In class, there was no division table like for other operations because the division frequently resulted in answers with a remainder or fraction. So, it would not be that helpful.

Chapter 7: Zero Tolerance

So, what about division by zero?

As noted, my elementary school teachers said there was essentially only one explanation to division by zero.

It was "undefined".

Not very satisfying.

They noted that if "defined", it would not be consistent with the rules of multiplication and its relationship with division. This just said to me that the mathematical system was lacking.

Traditional math *defines* the relationship between multiplication and division such that $(a \div b)$ or $a/b = c$ means that $a = (c \times b)$.

For example, $6/2 = 3$ implies that $6 = (3 \times 2)$.

It also *defines* that any number n, multiplied by zero is zero. And zero divided by any number is also zero.

That is, $(n \times 0) = 0$ and $0/n = 0$.

For example, $(3 \times 0) = 0$ and $0/3 = 0$.

Traditional math also says that any number n, divided by itself equals one. Except zero, since division by zero is "undefined" of course.

How convenient.

So, $3/3 = 1$ and $n/n = 1$ but *excluding* $n = 0$ or $0/0$.

The argument against division by zero based on the *defined* relationship between multiplication and division goes as follows:

Start with the equation $a/b = c$. By *definition*, this means that $a = (c \times b)$.

If $b = 0$, then $a = (c \times 0) = 0$, which means that c can have any value we want since $(n \times 0) = 0$. So, it makes no sense according to the argument.

However, I find this logic pretty circular and self-serving.

Stating the result another way, we can say that *defining* $b = 0$ implies that $a = 0$.

Afterall, it's already been *defined* that *any* number times zero equals zero.

In addition, it's *defined* that any number times one is itself. That is, $(n \times 1) = n$ so n can be any number.

What's the difference?

Another common, but more complicated "proof" follows. This argument tries to trick the system by "disguising" division by zero and shows that it still doesn't work.

Skip to the next chapter if too mind boggling.

It goes as follows.

Start with the equation $1 = a$ and multiply both sides by a so that $a = a^2$.

Now subtract 1 from both sides which gives you

$(a - 1) = (a^2 - 1)$. Next, divide both sides by $(a - 1)$ to get $(a - 1)/(a - 1) = (a^2 - 1)/(a - 1)$.

Since $a = 1$, we are technically dividing by zero, though the system doesn't "know" it.

Don't tell.

Again, this argument assumes that $n/n = 1$ for every number except zero. In this case $n = (a - 1)$ so

$(a - 1)/(a - 1) = 1$.

This means $1 = (a^2 - 1)/(a - 1)$.

We know from high school math that we can expand the numerator on the right side of the equation to the expression $(a - 1)(a + 1)$.

Further, expanding and simplifying gives the equation

$1 = (a - 1)(a + 1)/(a - 1)$ and $1 = (a + 1)$.

However, we started with $a = 1$, so substitution gives $1 = (1 + 1)$, and $1 = 2$.

Hold on here!

If we go back to the original equation of $a = a^2$ we easily see that there is a solution. In fact, there are *two* solutions under the accepted rules of traditional math.

Namely 1 and 0.

This just tells me that there's a problem with the system's rules of division.

Let's investigate this one step further and go back to the relationship between multiplication and division where $a/b = c$ which means that $a = (c \times b)$.

However, this time let's assume that $n/0 = 0$ and that $0/0 = 0$.

Let's use actual numbers to make it easier to follow. So $(2 \times 0) = 0$ and $(4 \times 0) = 0$.

That means $(2 \times 0) = (4 \times 0)$.

Dividing both sides by 0 gives $(2 \times 0)/0 = (4 \times 0)/0$, which transforms to $2 \times (0/0) = 4 \times (0/0)$.

However, this time it simplifies to $(2 \times 0) = (4 \times 0)$ or $0 = 0$.

It works!

Those interested can apply it to the general case for any number n, and find that it still works.

So, maybe, just maybe, the division operation as we know it, is not consistent with the *definition* that $n/n = 1$.

Note my emphasis on the words *define* and *definition* throughout this discussion. Remember that the foundations of traditional mathematics are based on *definitions* and *axioms* that you need to accept as "self-evident" truths without proof.

That really means they are dependent on your intuitive observation of nature. However, since the traditional mathematicians are in charge of this matter, it depends more specifically on *their* intuitive observation of nature.

Next is my treatment of division by zero.

See what you think.

Ø

Chapter 8: Division by Zero

Let's take the division process *literally*. It's a way of thinking about division for all numbers, not just zero.

So, I can apply the operation *division by zero* to all values. It is based on what you observe in nature.

If you have one item and you divide it by zero, you still have that one item since you didn't operate on it at all. You divided it zero *times* or divided it by nothing, so didn't change it.

So "1 *zed* 0" is one, or (1 Ø 0) = 1.

This holds for any number of items or values. So for all numbers *n*, (*n* Ø 0) = *n*.

Next, take one item and divide it by 1 or one *time*. The solution is 2 because you just divided that item once by cutting it in half. So, (1 Ø 1) = 2.

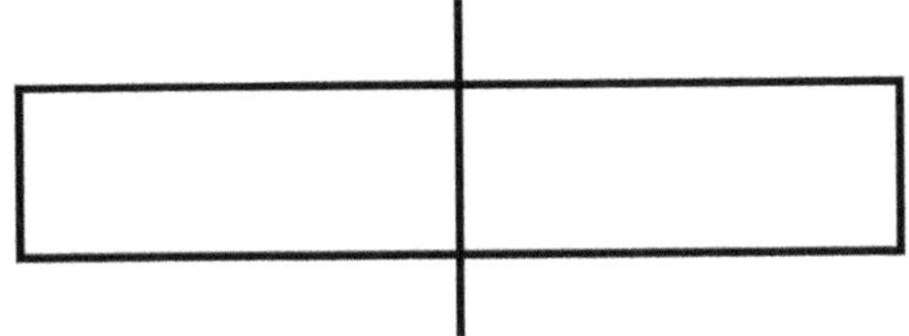

Along these lines, one item divided two times becomes (1 Ø 2) = 3.

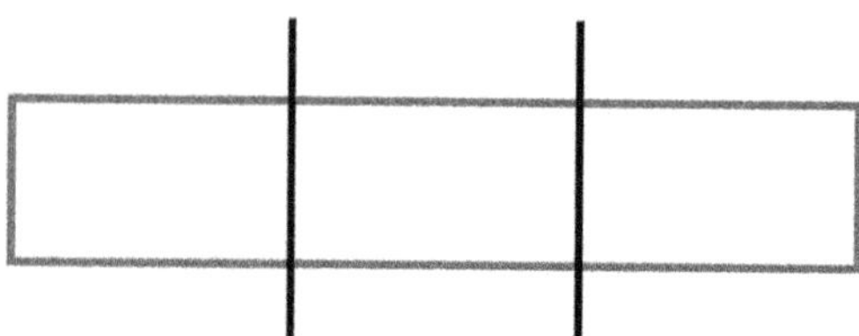

If you have two objects and divide them one time, then (2 Ø 1) = 4.

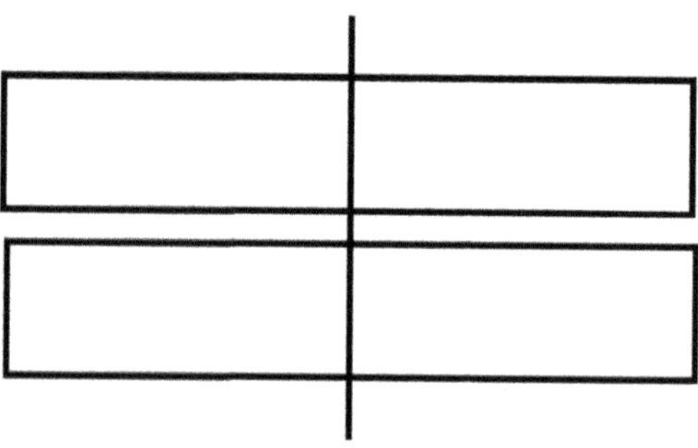

By the same token, two items divided 2 times under *division by zero*, gives you (2 Ø 2) = 6.

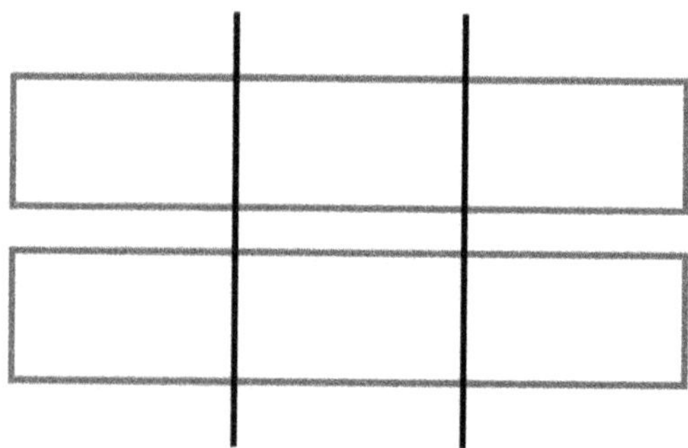

If you construct the table, it looks like this:

Ø	**0**	**1**	**2**	**3**	**4**	**5**	**6**	**7**	**8**	**9**
0	0	0	0	0	0	0	0	0	0	0
1	1	2	3	4	5	6	7	8	9	10
2	2	4	6	8	10	12	14	16	18	20
3	3	6	9	12	15	18	21	24	27	30
4	4	8	12	16	20	24	28	32	36	40
5	5	10	15	20	25	30	35	40	45	50
6	6	12	18	24	30	36	42	48	54	60
7	7	14	21	28	35	42	49	56	63	70
8	8	16	24	32	40	48	56	64	72	80
9	9	18	27	36	45	54	63	72	81	90

If you analyze the table carefully, you will find that $(A \text{ Ø } B) = A \times (B+1)$ when expressed within the system of traditional math.

I have zero comments at this point.

Chapter 9: Multiplication Revisited

Let's get back to multiplication.

As noted, we learned that multiplication is a shorthand notation for repeated addition. But what if we think about it like it occurs in nature. What if you took the act of multiplication literally or "naturally"?

That is, if you take one item and multiply or replicate it one time,

you end up with two items.

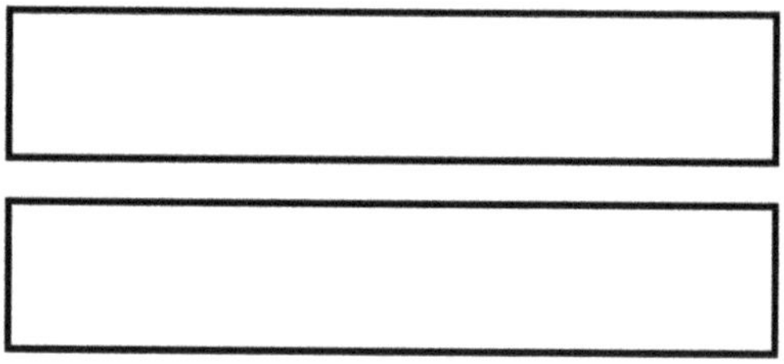

Start with just one item again,

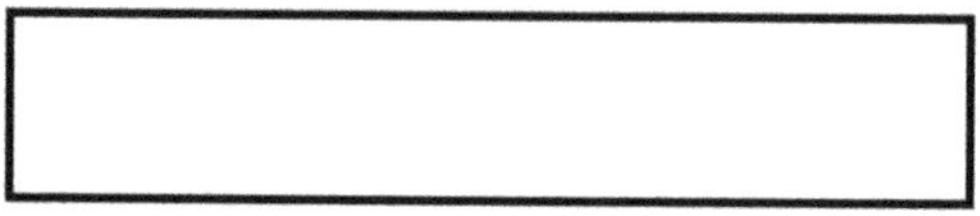

and multiply or replicate it *two* times. You get a total of three items.

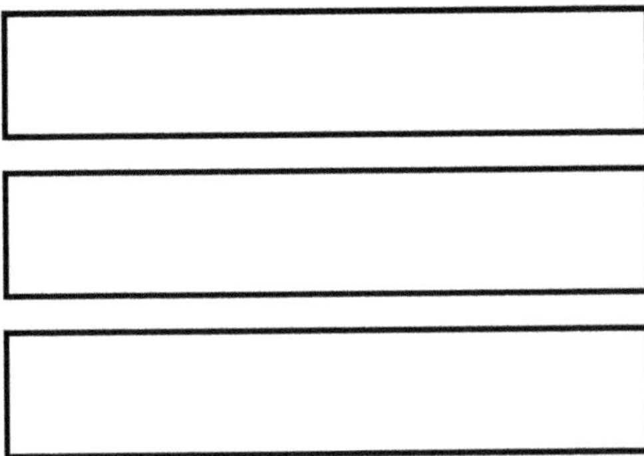

Let's call it *natural multiplication* and use a big, bold ***X*** as the operator symbol. The table looks like this.

X	**0**	**1**	**2**	**3**	**4**	**5**	**6**	**7**	**8**	**9**
0	0	0	0	0	0	0	0	0	0	0
1	1	2	3	4	5	6	7	8	9	10
2	2	4	6	8	10	12	14	16	18	20
3	3	6	9	12	15	18	21	24	27	30
4	4	8	12	16	20	24	28	32	36	40
5	5	10	15	20	25	30	35	40	45	50
6	6	12	18	24	30	36	42	48	54	60
7	7	14	21	28	35	42	49	56	63	70
8	8	16	24	32	40	48	56	64	72	80
9	9	18	27	36	45	54	63	72	81	90

Notice anything?
Check the table and you'll find that (*A* ***X*** *B*) = (*A* **Ø** *B*).
Yes, it's the same table as *division by zero*.

Could it be that in nature, division and multiplication are inherently the same thing?

Let's go back to basic biology class.

If one cell "multiplies" or divides, you end up with 2 cells. Same as (1 Ø 1) equals 2.

If you go back to the original cell you started with and prompt it to divide itself once more, you get a grand total of 3 cells. The original cell and the two replicated ones. Same as (1 Ø 2) equals 3.

Hmm…

Of course, my basic biology class claimed that when a cell divides repeatedly, the original and new cells simultaneously divide again and again. I never understood how one cell "knows" how to divide at the same time as the other one.

But I was never good at biology.

Let's continue to explore. But first we need to visit some additional definitions and concepts in the traditional mathematical system.

Some purport to be important descriptors of nature.

Chapter 10: Natural Numbers

Natural numbers should be simple.

However, some mathematicians have complicated them.

Some call them "counting numbers" and some call them "whole numbers".

Some include zero and some don't.

However, "counting" numbers vs "counting numbers" just means "reciting" them. And obviously zero is a whole number since it is not a fraction.

Consider the situation where there are no apples on a table and someone says, "Count the number of apples on the table," or asks, "How many apples are on the table?"

If there are none, you simply count or say, "Zero".

So, in the interest of *Natural Math* I use the definition that includes zero.

Therefore, *natural* numbers are:

0, 1, 2, 3, 4, 5...

Natural numbers are not just about counting things but are abstract entities representing discrete concepts.

If we see a large apple and a small one, we see them as two apples. We don't see one apple and a fraction of an apple.

They are the basic building blocks of arithmetic and how we first interacted with numbers as children.

In that sense, *natural* numbers are the guardians of natural physical concepts.

Chapter 11: Integers

Integers are defined as all the whole numbers. They include *positive* and *negative* numbers as well as zero.

$$\ldots -5, -4, -3, -2, -1, 0, 1, 2, 3, 4, 5 \ldots$$

They form the structure of the number line and its basic markers.

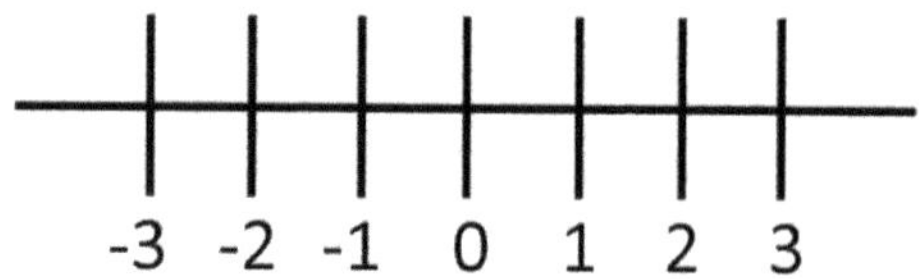

Ø

Chapter 12: Rational Numbers

Rational numbers have nothing to do with numbers acting rationally or logically.

It is simply a definition that mathematicians use to describe a particular collection of numbers with common properties.

Specifically, a *rational* number is one that can be expressed as a "ratio" of two whole numbers or *integers*.

Hence "ratio-nal".

This is typically defined as *a/b* where *a* is called the *numerator* and *b* the *denominator*.

So, all the *integers* are "rational" since they can all be expressed as a fraction or *ratio* with 1 as the *denominator*. For example, the fraction 3/1.

Fractions are also *rational* because they are already expressed as *ratios*. The most common ones being ½, ⅓, ⅔, ¾ etc.

You might think that fractions which have *denominators* that can be divided by 2 or 5 will easily convert into decimal notation since they seem to divide evenly into our *Base 10* number system.

For example:

1/5 = 0.20
¼ = 0.25
1/20 = 0.05

However, this is frequently not the case:

1/12 = 0.083333...
1/22 = 0.045454...
1/30 = 0.033333...

On the other hand, those with multiples or powers of two in the denominator seem to convert easily, as every subsequent power appears to add one more decimal place to the number and they all end in 5. For example:

½ = 0.5
¼ = 0.25
1/8 = 0.125
1/16 = 0.0625
1/32 = 0.03125
1/64 = 0.015625
1/128 = 0.0078125
1/256 = 0.00390625

And while many fractions have decimals that repeat indefinitely,

1/3 = 0.33333…

and,

2/3 = 0.66666…

others repeat in clusters.

1/11 = 0.09 09 09 09 09 09…

Some in long clusters:

1/7 = 0.142857 142857…

Some in really, really long clusters. Here's one with 22 repeating decimals:

1/23 = 0.0434782608695652173913
0434782608695652173913…

In fact, one of the properties of *rational* numbers is that their decimal representation comes to an end or repeats.

One interesting decimal conversion is a fraction with 9s in the *denominator*.

1/9 = 0.11111…
2/9 = 0.22222…
3/9 = 0.33333…
25/99 = 0.25252525…
311/999 = 0.311311311311...

Notice any pattern?

Chapter 13: Irrational Numbers

It only stands to reason that a number which is not rational is *irrational.*

Again, this does not mean that it is not logical.

Since a *rational* number is one that can be expressed as a ratio of two numbers, it follows that an *irrational* number cannot.

We noted that *rational* numbers can be identified by their finite or repeating decimals. Similarly, *irrational* numbers can be identified by decimals that don't repeat and go on forever.

So, what are they?

Well, the most well-known is the square root of 2, also expressed as $\sqrt{2}$.

The ancient Greek mathematician, Pythagoras, came across it when he proved that the sum of squares of the sides of a right triangle is equal to the square of its longer side or *hypotenuse*. Also expressed as $x^2 + y^2 = z^2$.

Remember that from high school geometry?

The easiest example is the triangle with sides 3-4-5. The math shows that $(3^2 + 4^2) = 5^2$ since $(9 + 16) = 25$.

To find $\sqrt{2}$, consider a triangle where each side is 1, then we have $(1 + 1) = 2 = z^2$.

So, z equals the square root of 2 or $\sqrt{2}$.

Expressed as a decimal, it is:

1.41421 35623 73095…

However, this sequence does not repeat.

One fractional approximation is:

99/70 = 1.41428...

Although √2 is an *irrational* number, I'd consider it "natural" in the sense that it is observable in nature and easily found.

Simply take a piece of string and fold it in half so you have two equal sides. Now, open up those halves so they are separated by a right angle. The third side connecting the ends or *hypotenuse* will be exactly √2.

Another method is to draw a square. Assign the value of one unit to each side. It doesn't matter whether you are using feet, meters, whatever. It just means that the sides of the square are equal.

Now draw a diagonal across it. That diagonal will be √2 relative to each side.

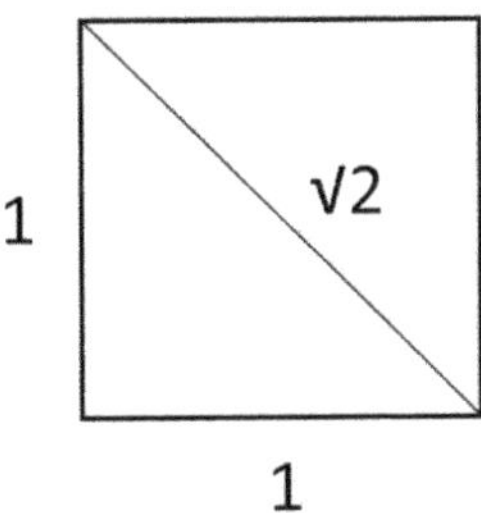

This brings us to the most famous *irrational* number of them all.

Pi.

Chapter 14: Pi

Pi is undoubtedly the most important mathematical concept in nature.

It is an *irrational* number symbolically represented by the Greek letter π.

It is *irrational* because it can't be expressed as a *ratio* of whole numbers or *integers*.

Pi is defined as the *ratio* of the circumference *C*, of a circle to its diameter *d*, or *C*/*d*. Don't ask me why a capital "C" is typically used. Maybe because it's the bigger value.

Importantly, the *ratio* defined by *pi* is constant regardless of the size of the circle. That means it not only defines a circle, but it defines the *concept* of a circle.

My schoolteachers expected us to know it to five decimal places.

3.14159

The first 10 are:

3.14159 26535

And the first 50 are:

3.14159 26535 89793 23846 26433
83279 50288 41971 69399 37510

Over the years, mathematicians tried to approximate it with fractions and found the following to be pretty good:

22/7 = 3.14285 71428...
333/106 = 3.14150 94339...
355/113 = 3.14159 29203...

However, they were frustrated that they couldn't attach an exact numerical value to it and had to satisfy themselves with a never-ending arithmetic series that converges to its exact value, though it never gets there. Mathematicians call this a *limit.*

A *limit* is defined in math as the boundary of a never-ending computation that gets closer and closer and closer to the boundary. The boundary is never reached, but it is also never exceeded.

Think of it like creating fractions with bigger and bigger *denominators* such that the fraction is $1/n$ and n progressively becomes bigger and bigger until your calculator runs out of decimal places.

Start with 1/2, then 1/10, then 1/1,000 then 1/1,000,000 etc. The fractions approach zero, but never reach it. So, zero might be called the *limit* of that process.

One non-ending series for computing *pi* is:

$$\pi/4 = + 1/1 - 1/3 + 1/5 - 1/7 + 1/9 - 1/11 + \dots$$

Or,

$$\pi = + 4/1 - 4/3 + 4/5 - 4/7 + 4/9 - 4/11 + \dots$$

Note that all the *denominators* are the counting series of odd numbers and the operations just alternate between addition and subtraction.

Although *pi* cannot be expressed as a number, the π symbol can substitute for it in the traditional mathematical system.

So, you don't need to know the actual number to use it. Simple examples might be:

$$\pi - \pi = 0$$
$$\pi \div \pi \text{ or } \pi/\pi = 1$$

Maybe these are too simple, but you get the idea.

Importantly, you can easily determine *pi* with a simple string exercise just like finding √2.

Here goes.

Take a piece of string and wrap it around a soda can. Cut it where it meets and you have the circumference of that circle.

Now, stretch that piece across the top of the can, which represents its diameter, and cut it at the edge. Repeat this until you have only a small piece left. You should be able to do this three times. The last little bit is the length of the fractional portion of *pi*.

Alternatively, you can just fold the originally cut string back upon itself twice which will result in three sections plus the fractional bit.

You can also construct a number line that pinpoints its location. Though this one requires a bit of dexterity.

Use a straight edge to draw a line on a piece of paper.

Take a coin or other disc-like object that can be rolled and mark one spot on the perimeter. A small, plastic vitamin bottle works well.

Roll it along the line until it gets back to itself. Mark the spot on the line. This represents the circumference of the disc.

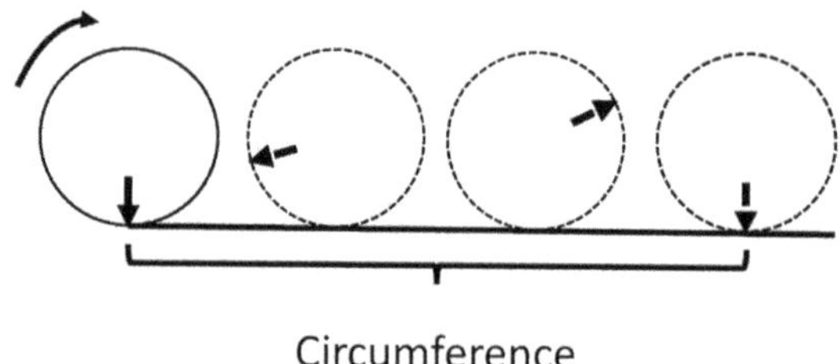

Circumference

Now, take the diameter of the coin using a compass.

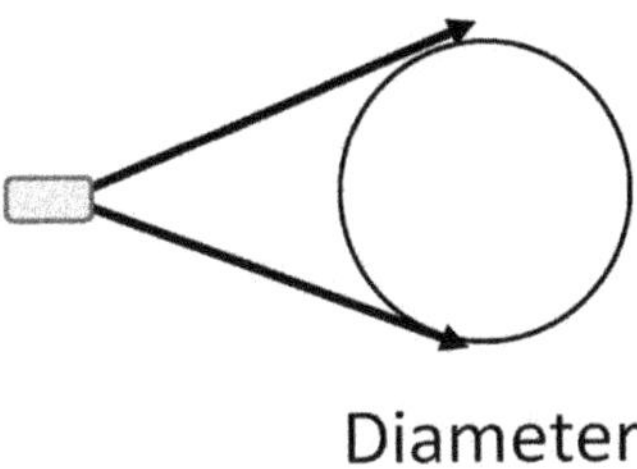

Finally, mark off three sections on the line using the diameter. The remaining section between the third mark and the circumference is the fractional or decimal portion of *pi*.

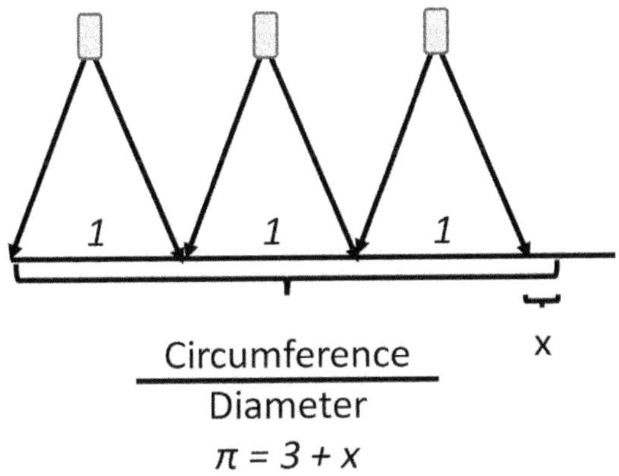

Many people just think of *pi* as a number.

However, it represents an important concept that occurs in nature.

As noted, *pi* is defined as the *ratio* of the circumference of a circle divided by its diameter. So, it's bizarre that it can't be expressed as a *ratio* of *natural* numbers. At the same time, it has basis in the natural world because circles are everywhere.

Mathematically, it typically occurs in expressions or functions that measure cyclical or periodic processes.

Cyclical or periodic events are easily observed in nature. The day, the orbit of planets, the seasons, the blooming of plants, the vibration of a string, harmonics in music, the oscillation of a radio wave and so forth.

Maybe life and death itself.

So importantly, *pi* is a constant in a world of change.

And now, it's officially on the calendar.

"Worldwide *Pi* Day".

March 14 or 3/14.

Chapter 15: Euler's Constant

Euler's Constant, represented by the symbol e, is associated with Swiss mathematician Leonhard Euler, pronounced "Oiler".

Like *pi*, it is one of the important numbers that relates math to nature.

Also like *pi*, e is an *irrational* number because it cannot be expressed as a fraction. It is also a constant.

However, it can be approximated with a non-ending series. One example is:

$$e = 1 + 1/(1 \times 1) + 1/(1 \times 2) + 1/(1 \times 2 \times 3) + 1/(1 \times 2 \times 3 \times 4) + 1/(1 \times 2 \times 3 \times 4 \times 5) + \ldots$$

Mathematicians have a notation for the sequence of multiplications as expressed in the *denominator*. For any number n, they call it *n factorial* and represent it as $n!$

It is simply the product of any number and all the *natural* numbers preceding it, excluding zero. For example, *4 factorial* is:

$$4! = (1 \times 2 \times 3 \times 4)$$

Interestingly, they *define* $0! = 1$.

Whoops!

That's also the value of 1!

Another *definition* of convenience.

Anyway, in the simplified notation,

$$e = 1/0! + 1/1! + 1/2! + 1/3! + 1/4! + 1/5! + \ldots$$

Its approximate value to 15 decimal places is:

2.71828 18284 59045...

Some fractional approximations are:

$$87/32 = 2.71875$$
$$878/323 = 2.71826\ 62538\ldots$$

In the natural world, *e* is mostly associated with growth and decay. It is the base rate of growth where things grow or decay continuously and proportionally. This includes populations, bacteria, nuclear radiation as well as something as mundane as bank interest. Notably, *compound interest.*

Let's explore *compound interest* since most of us know about it in the context of our bank savings account.

You make a bank deposit and they pay you interest linked with a defined annual percentage rate. The number of times they calculate it within a year refers to the number of times they *compound* it. Typically, on a yearly, quarterly or monthly basis.

If you don't withdraw the interest, it becomes part of the *principal,* or starting amount, the next time that round of interest is calculated.

Let's start with a *principal* of $100 and the simplest example in which they only *compound* it annually, or one time at the end of a year.

So, if you deposit $100 at 5% interest *compounded annually*, you have $\$100 \times (1 + 5\%)$. Alternatively, this is $100 \times (1 + 5/100)$ or $\$100 \times (1.05) = \$105$ after one year.

If you keep it all in the bank without withdrawing the interest, then the next year around means $\$105 \times (1.05)$ or $100 \times (1.05)^2 = \$110$. Rounded up to the nearest dollar.

Keeping all the money in the bank for a third year gives $\$110 \times (1.05)$ or $100 \times (1.05)^3 = \$116$.

However, some banks offer to *compound* the interest more frequently. Maybe monthly, weekly or even daily.

Sounds good.

However, you need to read the fine print. You'll note that they still calculate or *compound* the interest on the basis of an *annual* rate.

What's that mean?

Getting too deep into the math gets very complicated.

Suffice to say that the more frequently you *compound* the interest, the smaller the incremental amounts become. And there's a *limit* to the number you can reach.

The formula for calculating the *compound interest* for an annual interest rate *r*, and *n* number of times in one year is:

$$(1+r/n)^n$$

So, if you start with amount, A_0 you end up with amount A_1 at the end of the year.

$$A_1 = A_0 \times (1+r/n)^n$$

Beginning with \$100 at an annual interest of 5% for *n* *compounding* periods in one year, you get the following:

Annually = \$105.00000

Monthly = \$105.11619

Weekly = \$105.12458

Daily = 105.12675

Hourly = 105.12709

Minute-by-Minute = 105.12711

You easily see from the below calculations that the *incremental* amount gets increasingly smaller.

Monthly vs Annually	0.11619
Weekly vs Monthly	0.00839
Daily vs Weekly	0.00217
Hourly vs Daily	0.00034
Minute vs Hourly	0.00002

In fact, if you computed it *continuously*, it would reach a *limit*. That's were *e* comes in.

The previous formula,

$$(1+r/n)^n$$

approaches a *limit* as *n* goes to infinity. It is designated:

$$\lim_{n \to \infty} (1+r/n)^n = e^r$$

So, the maximum amount you can get from the bank under *continuous compounding* in one year is:

$$A_1 = A_0 \text{ x } e^r$$

Let's try and compare.

Below are the previous calculations, but stretched to 10 decimal places.

Annually	105.00000 00000
Monthly	105.11618 97882
Weekly	105.12458 41927
Daily	105.12674 96467
Hourly	105.12709 46366
Minute-by-Minute	105.12710 93874

Applying the above *limit* formula for *e* with $r = 5\%$ to 10 decimals gives us:

$$A_1 = 105.12710\ 96376$$

For my money, they can stop at daily *compounding*. Even if I deposit one million dollars, the difference between daily and minute-by-minute *compounding* only gives me an extra \$360 that I've likely spent beforehand.

The concept of *e* as a *limit* for this calculation can be visualized as:

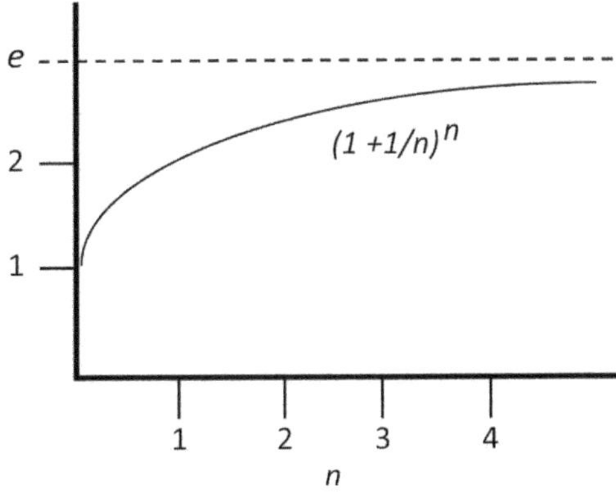

Mathematicians have shown that we can expand this over a number of time periods where *time* = t and:

$$A_1 = A_0 \times e^{rt}$$

So, to compute how much our \$100 would grow to over three years *compounded* continuously, we'd insert $r = 5\%$ and $t = 3$ and get,

$$A_1 = 100 \times 1.1618 = \$116.18$$

Let's show how this might work in the real world using a town with a population of 100 people that has demonstrated an estimated 5% population growth a year and see how many people might reside there in 10 years.

Assuming the same continuous growth rate for the 10 years, we'd set the starting population at $A_0 = 100$, *rate of growth* = 5% and $t = 10$ years.

So, $(r \times t) = (0.05 \times 10) = 0.5$.

Computing e to the power of 0.5 and multiplying times 100 gives a projected population of $A_1 = 165$ people.

Of course, when applying math formulas, we always need to note the *assumptions*. This example *assumes* constant growth of 5%. And a *constant* growth rate *assumes* a whole bunch of underlying dynamics. How many people left the town? How many arrived? How many died? How many mated and gave birth?

Be careful what you *assume*.

Chapter 16: Real Numbers

Real numbers comprise all of the above.

They are all the numbers that appear somewhere along a number line. As such, they include zero, *negative* numbers, *positive* numbers and everything in between including *rational* and *irrational* numbers.

I would contend that *negative* numbers should not be included because they are not really "real". As previously noted, I dare someone to show me a negative object. Natural or man-made.

Unreal.

Chapter 17: Imaginary Numbers

Well, it had to happen. Once traditional mathematicians departed from nature and introduced some new concepts, they should have expected that *unnatural* things would show up, despite their carefully defined and constructed system with all its rules.

The trouble started by introducing *negative* numbers.

As noted, there is no such thing as a *negative* number in either the natural or manmade world. You don't have a negative rock any more than you have a negative TV.

Then it got more complicated when mathematicians said that, *by definition*, multiplying a *negative* number by a *positive* number results in a *negative* number. So:

$-4 \times 3 = -12$.

And multiplying any two *negative* numbers results in a *positive* number. For example:

$-1 \times -1 = 1$.

$-3 \times -2 = 6$.

Does this mean that two wrongs make a right?

And what happens when you take the square root of a *negative* number? Say –1?

I bet they didn't see this coming when they defined multiplication for *negative* numbers.

They were stuck. So, instead of trying to figure it out, they just created the symbol *i* to represent the *concept* of whatever that solution is supposed to be. That is, an italicized, lowercase letter *i*. So:

$i = \sqrt{-1}$ and $i^2 = -1$.

Very imaginative.

Chapter 18: Reality Check

Now that we've introduced *real* and *imaginary* numbers, it's interesting to note that *Euler's Constant e* connects numbers across real and imaginary worlds.

It states that:

$$e^{\pi i} + 1 = 0$$

Imagine that!

Chapter 19: Fibonacci Sequence

Fibonacci is most famous for the numerical sequence named after him. The *Fibonacci Sequence.*

According to the story, the sequence was inspired by rabbit breeding habits. I've read the accounts, but not sure how a busy clerk in a customs house, traveling throughout the Muslim world and preoccupied with business calculations, found the time to breed and observe rabbits.

His famous book *Liber Abaci* devotes only one of several hundred pages to it. The rest are preoccupied with more practical problems related to buying and selling, computing shares of things merchants might need to divvy up, as well as their profits.

In short, the *Fibonacci Sequence* is allegedly based on the number of pairs of rabbits in the current month compared to the pairs of rabbits in the previous months.

It goes something like this.

Start with two rabbits. Different genders of course.

If every month they produce a pair which then produces a pair the following month, the total number of pairs of rabbits in each month should be:

1, 1, 2, 3, 5, 8, 13, 21, 34, 55,
89,144, 233, 377, 610, 987…

Any number in the sequence is just the sum of the previous two. So:

$1 + 1 = 2$

$2 + 1 = 3$

$3 + 2 = 5$

$5 + 3 = 8$

$8 + 5 = 13$

Of course, there's a big "if" in this. For one thing, it assumes that every month newly born rabbits are of opposite genders, feel like mating and give birth to another pair.

I guess it counts on the adage "breed like rabbits".

However, there's another problem. I'm no historian, but I do check some things.

In *Liber Abaci*, Fibonacci published the sequence as:

1, 2, 3, 5, 8, 13, 21, 34, 55, 89, 144, 233, 377…

Notice anything?

There's a "1" missing at the beginning.

In fact, Fibonacci described the initial state of things as the "beginning month" with one pair of rabbits and designated it "1". The "first" month was 2, the second 3, the third 5, the fourth 8, and so on, up to the twelfth or "ending month" which was 377.

It seems that mathematicians subsequently edited the sequence to make it fit with the definition that every number is the sum of the previous two.

Some literature starts the sequence with zero:

0, 1, 1, 2, 3, 5, 8, 13, 21, 34, 55, 89, 144, 233, 377…

I suppose this makes the *definition* more *definitive* because it can claim the sequence is based on the *natural* numbers. This makes zero the starting point which gives us (0+1) = 1.

Meanwhile, some claim that Hindu mathematicians already published the sequence without any reference to rabbits and accuse him of "fibbing" about it.

In Fibonacci's defense, I must say that his *Liber Abaci* does not present this sequence as an original discovery, but merely as an observation. Also, he does not present it as something that describes some universal phenomenon of nature.

So, I find problems with those who *do* claim that it *does* describe nature.

Notably, some claim that it appears throughout the plant world.

Examples include the pattern of spirals that appear in sunflowers, pine cones, pineapples as well as the position of leaves as they spiral up the stem of some flowers and the number of peas in a pod.

I thought I'd see for myself.

I checked the number of peas in a pod after a quick grocery store purchase.

The first pod contained five and the next eight.

So far, so good.

However, subsequent pods produced mostly six and ten peas.

I also checked out a pineapple. Reportedly, the pattern of geometric scales that spiral around each other are *Fibonacci* numbers.

I couldn't figure it out.

Fibonacci numbers reportedly also appear in artichokes. Unfortunately, my grocery store was sold out.

Oh, well.

In the insect world, the *ratio* of the number of female honey bees to male honey bees in any hive is reportedly a *Fibonacci* number.

I'll take that one on faith.

Despite my failure to find *Fibonacci* numbers in nature, I find it beautifully simple, easy to memorize and comprised exclusively of *natural* numbers.

And speaking of beauty, there's a mathematical phenomenon that reportedly *quantifies* our concept of natural beauty.

The *Golden Ratio.*

Ø

Chapter 20: The Golden Ratio

The *Golden Ratio* is also called the *Golden Proportion,* the *Golden Mean*, the *Golden Section*, the *Divine Ratio* and the *Divine Proportion.*

Reportedly, the ancient Greek mathematician and father of geometry, Euclid, proposed it as the *ratio* that is most beautiful or pleasing to the eye.

It is defined as the *ratio* of two quantities a and b where $a > b$, such that $a/b = (a + b)/a$ and represented by the Greek letter *phi* or φ.

You can verbalize it as follows:

"The *Golden Ratio* is the *ratio* of two numbers such that the *ratio* of a large number to a smaller number is the same *ratio* as the sum of both numbers to the larger number."

Phi to 15 decimal places is:

1.61803 39887 49894…

Geometrically, it can be represented as two attached line segments:

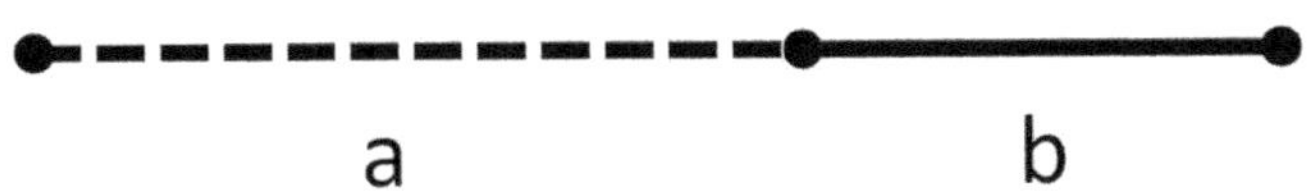

It can also be depicted as a rectangle whose sides represent the *ratio*, called a *Golden Rectangle.*

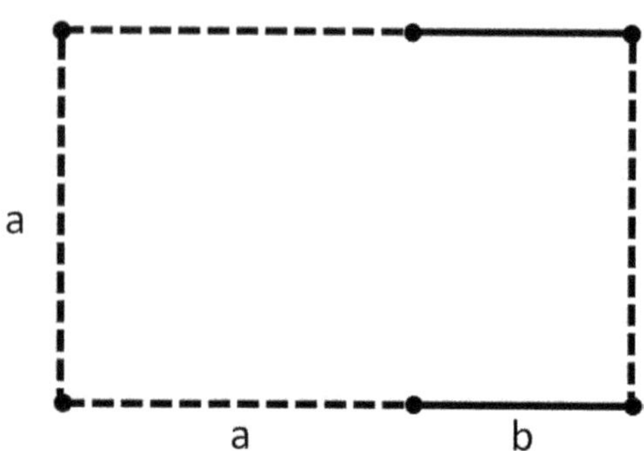

Pleasing to the eye? What about compared to other proportions?

See what you think. Below are rectangles of various proportions both horizontal and vertical. Pick the one you find most pleasing, then check *Appendix B* to see how it compares to a *Golden Rectangle*.

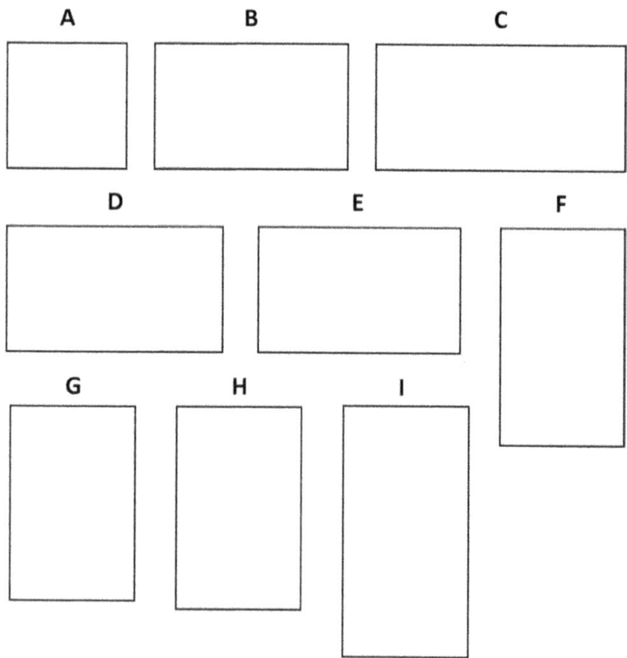

Meanwhile, what if you defined it as the *ratio* of the sum of both numbers to the *smaller* number? This would mean the relationship is $a/b = (a + b)/b$. Algebraically, this is more symmetrical since both denominators are b. And symmetrical things seem to be more visually pleasing.

Presented as a rectangle, you get:

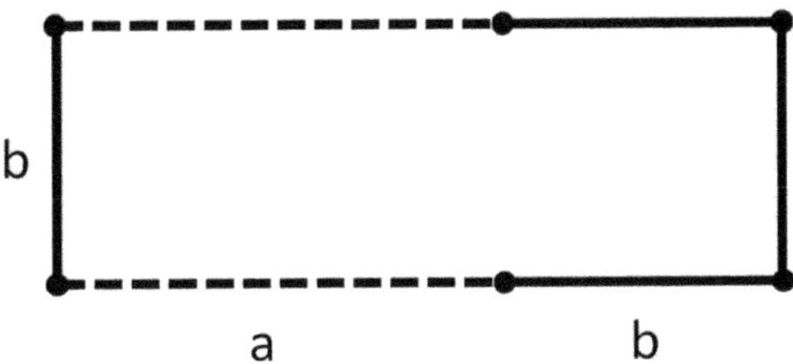

More or less pleasing?

As they say, beauty is in the eye of the beholder.

Hah!

I confess.

I constructed this rectangle using the lengths of a and b from the previous diagram. However, that doesn't mean that the *ratio* holds.

In fact it doesn't.

If you take the equation $a/b = (a + b)/b$ and multiply both sides by b, you get $a = (a + b)$ *or* $b = 0$.

Obviously, that means the width is zero, so there is no rectangle that satisfies this relationship.

Let's consider the search for *phi* in nature.

Many claim that it appears throughout the man-made world.

Various studies have explored the proportions of artists' paintings as well as the proportions of subjects within paintings looking to find *phi*. Others have researched ancient and modern architecture in search of the *Golden Proportion*. Most notably, the dimensions of the Greek Parthenon. However, it seems that *phi* was more often forced into them than found there.

Independently, I measured my TV and computer screens, different decks of playing cards, some of my favorite paintings, plus international and US typing paper standards. Respectively, A4 and 8½ x 11. None of them support the *Golden Ratio*.

However, the dimensions of my book covers, which are 8 x 5, come close at 1.60.

So, buy more of my books and frame them.

Mathematically, the value of *phi* can be expressed as:

$$\varphi = (1 + \sqrt{5})/2 = 1.61803\ldots$$

Calculating this to 15 decimals gives:

1.61803 39887 49894

And it has some interesting mathematical properties. Notably:

$$\varphi = 1 + 1/\varphi \text{ and } \varphi^2 = \varphi + 1$$

Very mysterious, if not outright mystical.

You'll note that, like π, the symbol can be used as a mathematical quantity without knowing the exact value of the number it represents.

Mathematicians found the value $(1 + \sqrt{5})/2$ using basic algebra as follows:

Start with the definition $a/b = (a + b)/a$.

But let's use the notation x and y to avoid confusion as we do the math. You'll see why in a minute.

So, $x/y= (x + y)/x$.

Assume $y = 1$, then $x/1 = (x + 1)/x$.

Multiplying both sides by x transforms to $x^2 = (x + 1)$ or the *quadratic* equation $(x^2 - x - 1) = 0$. Recall from high school algebra that a *quadratic* equation has the format $(ax^2 + bx + c) = 0$ with two solutions defined by the equation:

$$x = \frac{-b \pm \sqrt{b^2 - 4ac}}{2a}$$

This is the standard notation for the equation, so you see why I changed (a, b) to (x, y).

But *two* solutions?

Yep!

We learned that you're supposed to pick the one that makes the most sense for solving your problem and reject the other.

Interesting that mathematicians unabashedly say the correct solution requires a *judgement* call.

Often one solution is a *negative* number. You typically reject that one since there are no negative objects in the real world.

Hmm…

Anyway, one solution of this equation is $(1 + \sqrt{5})/2$ or *phi*. The other is the $(1 - \sqrt{5})/2$ which is about –0.61803.

This also happens to be the *negative inverse* of *phi* or the value $-2/(1 + \sqrt{5})$.

That's interesting.

It's also interesting that both solutions seem to have the same decimal expansion. At least to 10 places:

1.61803 39887
-0.61803 39887

Now, let's go back to the mathematical definition of *phi* whereby $a/b = (a + b)/a$.

Expanding the math gives $a/b = (a/a) + (b/a)$.

But, $a/b = \varphi$, $(a/a) = 1$ and $(b/a) = 1/\varphi$. Substituting symbols and values brings us back to:

$$\varphi = 1 + 1/\varphi$$

And, if we multiply both sides by φ, we get:

$$\varphi^2 = \varphi + 1$$

Not so mysterious afterall!

But wait a minute. They said *assume* $b = 1$.

Who decided that?

Instead, let's assume that $b = 2$.

This gives us $a/2 = (a + 2)/a$. Now multiply both sides by $2a$ to simplify the equation to $a^2 = 2(a + 2)$ which transforms to $(a^2 - 2a - 4) = 0$.

This produces two solutions which don't seem to be related to anything.

Specifically:

3.23607 and –1.23607

However, if you divide both solutions by 2, you get 1.16803 and –0.61803 which are the two solutions to *phi's quadratic* equation.

This makes sense since the definition of *phi* is based on *ratios*. So, in order to maintain the *ratio* a/b, we also need to multiply a times two if we multiply b times two.

One interesting property of *phi* is that is it is related to the *Fibonacci Sequence*.

Specifically, if you divide a *Fibonacci* number by its predecessor in the sequence, the *quotient* converges to *phi*. So, the bigger the number in the sequence you start with, the closer the approximation to *phi*.

Consider the *Fibonacci Sequence* again:

1, 1, 2, 3, 5, 8, 13, 21, 34, 55,
89, 144, 233, 377, 610, 987…

We find that,

$$8 \div 5 = 1.6000$$

While,

$$13 \div 8 = 1.62500$$

And,

$$21 \div 13 = 1.61538\ldots$$

Going out a bit further gives you,

$$987 \div 610 = 1.61803\ldots$$

which is already the value of *phi* to five decimals. An easy way to find *phi* naturally is to draw an equal-sided *equilateral* triangle inside a circle. Now, draw a line through the midpoints of the triangle's two sides to the edge of the circle. The line segments a and b, will satisfy the condition $a/b = (a + b)/a$ and represent *phi.*

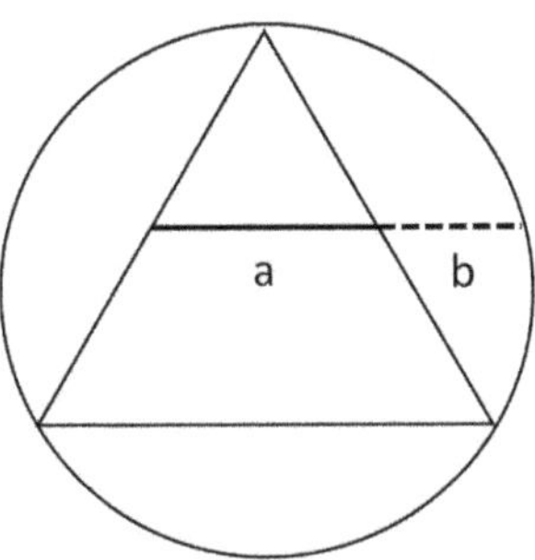

Meanwhile, it turns out that there is a geometric symbol which appears everywhere that totally embodies *phi.*

The five-pointed star, also known as a *pentagram.*

In elementary school, we learned how to draw it without taking our pencil off the paper.

We also learned that poking our fingers in the sand produced five points that could be connected by lines to produce the star with the middle finger making the top point.

The *pentagram* appears throughout history as a symbol in various faiths, orders and secret societies which some consider sacred or possessing magical powers.

Others claim that it represents the fundamental elements of the cosmos as defined by the ancients. Notably, *earth, water*, *fire*, *air* and *ether.*

Additionally, it might represent the five senses: sight, hearing, smell, taste and touch. As such, it symbolizes the unification of all things.

When turned upside down, some claim it symbolizes evil.

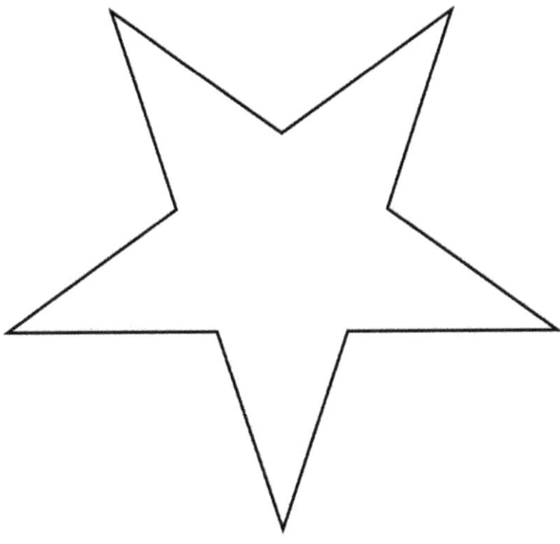

You can also inscribe a five-sided shape or *pentagon* with equal sides both inside and out.

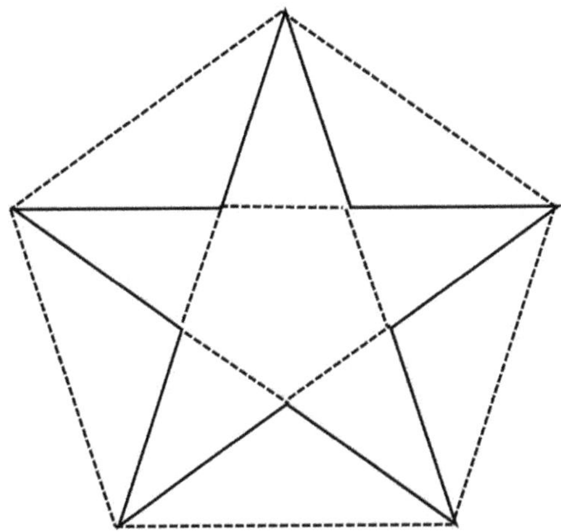

A *pentagram* also depicts the five limbs of man. My rendering below:

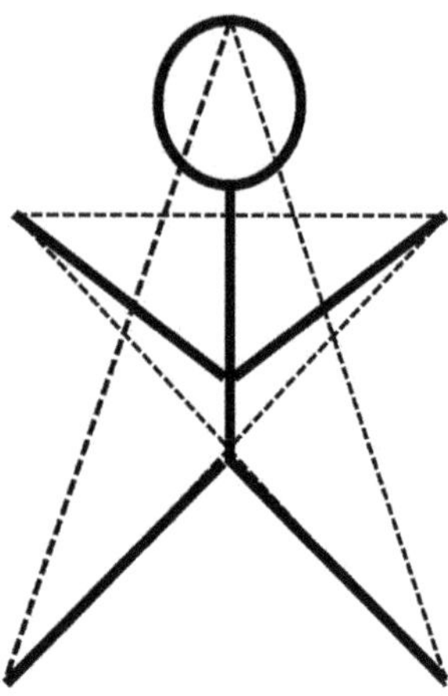

The pentagram is related to *phi* in that the *ratios* of the various segments are equal to it.

Notably, a/b = b/c = c/d = φ.

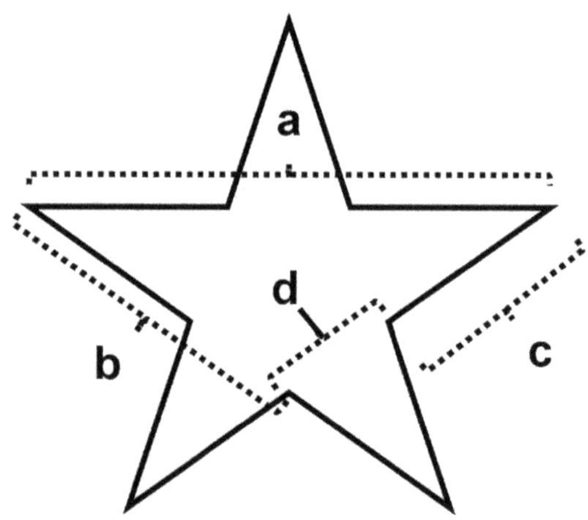

Importantly, these *ratios* are true no matter the size of the *pentagram*, just like *pi* with respect to a circle.

Since the *pentagram* embodies the definition of *phi*, it might also be considered a symbol of beauty.

Examples of the *pentagram* found in nature include the arrangement of the seeds of an apple core and the shape of a star fruit once they are cut open.

I cut open a number of apples to check it out and most of the time there were five seeds arranged as a *pentagram*.

Eve must have held the apple upside down when handing it to Adam.

Cutting open the star fruit results in three five-point stars. The outer shape of the fruit. Another star outlined inside and five seeds embedded within that star.

That's what I call star power.

Some claim slicing a banana reveals the same arrangement, but I found that a bit slippery.

Others claim that the human body also embodies *phi*. Specifically, the *ratio* of the following lengths:

$$\frac{\text{Shoulder to Fingertips}}{\text{Elbow to Fingertips}} = \frac{\text{Hip to Ground}}{\text{Knee to Ground}} = \varphi$$

I measured mine, but they are not even close.

Maybe I'm not human.

But what really, really bothers me is that the *Golden Proportion,* which mathematicians typically *define* as the embodiment of beauty in nature, is an *irrational* number. Again, it is a number that cannot be expressed as the ratio of two *natural* numbers.

Totally irrational.

Chapter 21: Spirals

Spirals are found throughout nature.

Seashells, galaxies, plants, fruit, tornadoes, hurricanes, whirlpools, human *DNA*, flushed water swirling down a toilet bowl. And the grooves on a vinyl record. As in "record player" or *phonograph*. Ever hear of that?

A *spiral* is just a curve that expands away from a central point as it rotates outwards. The type of *spiral* depends on how the curve's formula is mathematically defined.

The math gets a bit interesting. But in general, the mathematical formula defines a relationship between the *radius* r, or length of a line emanating from a center point and its angle θ as the angle gets bigger relative to a base line.

Whew!

It's easier to visualize rather than explain. You see that the radius r, gets longer as the angle θ gets bigger:

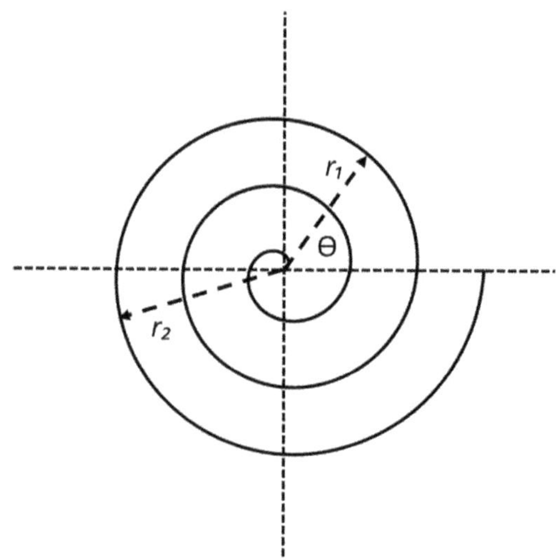

You can also have a *spiral* in three dimensions. It's just like pulling open a spring and called a *helix*. In order to keep it simple, let's just focus on some examples of two-dimensional *spirals*.

The simplest is an *arithmetic spiral* whereby the space between each successive point on the curve or turn is a constant.

Now consider a *spiral* where the *radius* length is determined by a power of *e*. This is called a *logarithmic spiral* and often found approximated in nature.

You can also construct a spiral using the *Golden Ratio* as well as the *Fibonacci Sequence*. In fact, they are close approximations of each other. However, the *spiral* closest to things in nature is the *logarithmic spiral*.

Sorry, Leonardo.

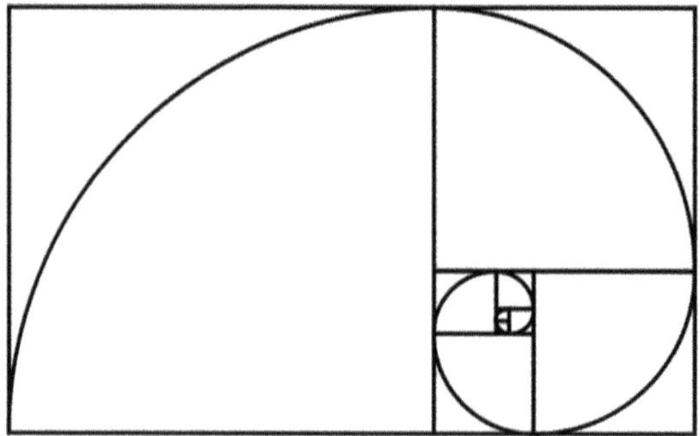

Separately, here's a thought I had in elementary school when we first learned about astronomy.

At that time, my teachers claimed that the stars in our *Milky Way* galaxy were so far away that you'd have to travel much faster than the speed of light to reach them in a reasonable amount of time. The distance is measured by the number of years that light takes to cover that distance. *Light years*.

They also claimed that traveling anywhere close to the speed of light, let alone faster than it is impossible.

We also learned that our *Milky Way* galaxy is a *spiral* galaxy.

I suggested that maybe the stars just *appear* to be very far away because light travels around the *spiral* path of our galaxy and by the time it reaches us, it has covered *light years* of distance.

However, what if the stars are just located along one of the next curves of the *spiral*, so really very close. We just need to figure out how to jump from one curve to the next rather than travel around the *spiral* of space. Just like an energized *electron* in an atom jumps from one state of energy to another in a *quantum leap*.

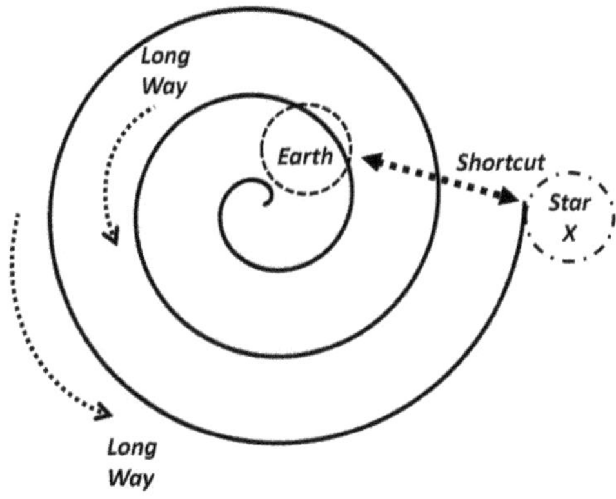

This was intuitively obvious to me. It was no different than climbing over the chain-link fence in my backyard to get to a friend's house on the other side of the block. It was merely a shortcut.

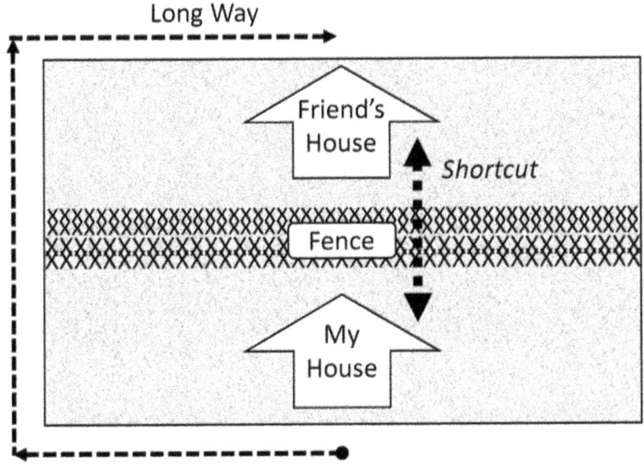

Of course, this was harder than strolling around the block. You needed to energize yourself to get over the fence in one big jump so you didn't catch your clothes on the barbed top.

Sort of a personal *quantum leap*.

Chapter 22: Sets

A *set* is a simply a collection of items.

It could be a collection of fruit, people or other objects.

Even numbers.

In traditional math, a set is represented by placing the items in closed, curly brackets. The items within the *set* are called *elements*.

Many of the items we already described would be considered a *set*.

The set of *natural* numbers is expressed as:

$$\boldsymbol{N} = \{0, 1, 2, 3, 4, 5\ldots\}$$

And the set of *integers* as:

$$\boldsymbol{Z} = \{\ldots-5, -4, -3, -2, -1, 0, 1, 2, 3, 4, 5\ldots\}$$

Mathematicians use the bolded letter $\boldsymbol{Z}$ to signify the set of *integers* rather than $\boldsymbol{I}$ following the tradition of German mathematicians. It is the first letter of *zahlen* which means "numbers" in German.

These *sets* happen to have some properties in common based on their definitions.

However, a *set* can also be a collection of random numbers. Let's define such a *set* and label it $\boldsymbol{T}$ for *temere* which means "random" Latin.

$$\boldsymbol{T} = \{1, 5, 8, 11, 19\}$$

No big deal.

Sets included in other *sets* are called *subsets*. So, *integers*, *natural*, *rational* and *irrational* numbers are all *subsets* of *real* numbers.

As noted, *sets* don't need to be comprised of numbers.

Consider a *set*, ***F***, defined as fruit and the *set* ***A***, defined as apples.

Then ***A*** would be a *subset* of ***F***, since apples are contained within the collection of fruit.

If you have one *set* consisting of apples, grapes and bananas, and another *set* consisting of apples grapes and melons, then a *subset* consisting of apples and grapes would be called an *intersecting* or *overlapping set* since they are contained in both.

Graphically, it can be represented as a *Venn* diagram, named after mathematician John Venn as:

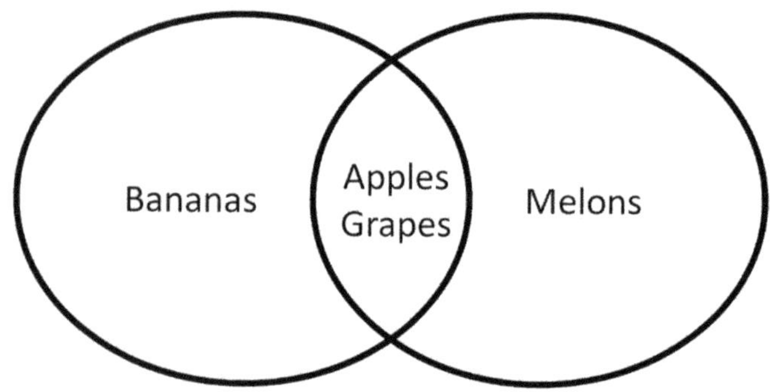

There is a whole body of mathematics that analyzes *sets* called *Set Theory* but it is beyond the scope of this book.

However, I introduce the idea because it helps discuss how collections or *sets* of numbers interact which is discussed shortly.

So, this is a good "*set* up" for the next chapter.

Chapter 23: Infinity & Beyond

Infinity means "endlessness" in most people's minds.

It is symbolized by ∞.

A figure eight with an apparent twist.

Aficionados call the symbol a *lemniscate* which is derived from the Latin *lemniscatus* meaning "decorated with ribbons" or the Greek ληµνίσκος *or lemniskos* meaning "ribbon" or more specifically, "wool ribbon". Presumably most ribbons were made from wool in ancient Greece.

We can consider infinity as the endlessness of linear distance which can go on forever, like outer space.

And we can also consider it as the endlessness of time. That is, infinity is a process or an action that continues without stopping.

It is a duality.

As such, it can also be symbolized by a circle.

You can go round and round and round a circle forever while covering a never-ending distance.

Aside from a circle, how else can we experience infinity in the natural world?

In geometry, it's accepted that there are an infinite number of points between any two points.

So, if I start with two points:

Then connect them with a line:

I'll have passed through an infinite number of points.

Right?

Does that mean I just did something infinite?

You decide.

One can also easily perform something "infinite" in three dimensions by constructing a *möbius strip*.

Simply take a ribbon of paper, twist it once and glue or tape the ends together. You have a *möbius strip*. You'll also note that it looks like the symbol for infinity ∞.

Now take a pen and trace along its surface. The surface is two-dimensional but your pen is traveling in three dimensions. And like traveling around the perimeter of a circle, you can continue forever.

It represents endlessness in both space and time.

Various number *sets* can also be considered infinite.

Take *natural* numbers.

You can count them until you drop.

We can also *prove* that they are infinite as follows.

Let's call n the absolutely largest number you can imagine. However, for every n you can think of, there is always one bigger or $(n + 1)$.

Not too difficult.

But when you plug *infinity* into arithmetic, it has more problems than zero.

Simply ask yourself the following.

What is $(\infty + 1)$?

What is $(\infty + 2)$?

Are they ∞ or something bigger?

Shouldn't they be bigger?

Is $(\infty + 2) > (\infty + 1)$?

Does $(\infty - \infty) = 0$?

Does $(\infty \times 2) = (\infty + \infty)$?

Does $(\infty \times 2) = \infty$?

And what about $(\infty \times \infty)$?

Some suggest that they all equal ∞. Others claim that they are all undefined. Let's see for ourselves.

First try addition using the number 2.

Start with $(\infty + 2) = \infty$, then subtract ∞ from both sides, giving $(\infty + 2) - \infty = (\infty - \infty)$.

If we assume $(\infty - \infty) = 0$, then $2 = 0$.

That also means for any number n, $n = 0$.

Not good.

Since it doesn't work, does this mean I just proved that the expression $(\infty - \infty) \neq 0$?

What a genius I am!

But if it doesn't equal zero, then what does it equal?

Let's try something else.

It's accepted that for any number n, $(n/n) = 1$, except for 0, of course.

So, let's assume that $(\infty/\infty) = 1$.

Now, consider $(\infty \times 2) = (\infty + \infty)$.

Dividing both sides by ∞ gives us the expression

$(\infty \times 2)/\infty = (\infty + \infty)/\infty$.

This transforms to $(\infty/\infty) \times 2 = (\infty/\infty) + (\infty/\infty)$.

Then, $(1) \times 2 = (1) + (1)$.

So, $2 = 2$.

Looks pretty good so far.

However, what happens when we try $(\infty \times 2) = \infty$?

Dividing both sides by ∞ gives $(\infty \times 2)/\infty = (\infty/\infty)$.

This transforms to $(\infty/\infty) \times 2 = \infty/\infty$.

This gives $(1 \times 2) = 1$ or $2 = 1$.

Oh well.

Let's take it one step further and now assume that the expression $\infty/\infty = 0$.

Consider $(\infty \times 2) = \infty$.

Dividing both sides by ∞ gives $(\infty \times 2)/\infty = (\infty/\infty)$.

This transforms to $(\infty/\infty) \times 2 = (\infty/\infty)$ or,

$(0 \times 2) = 0$ and $0 = 0$.

It works!

Let's keep going.

Now see if $(\infty \times 2) = (\infty + \infty)$ under the assumption that $\infty/\infty = 0$.

Dividing both sides by ∞ gives us the expression

$$(\infty \times 2)/\infty = (\infty + \infty)/\infty.$$

This transforms to $(\infty/\infty) \times 2 = (\infty/\infty) + (\infty/\infty)$.

Then $(0) \times 2 = (0 + 0) = 0$ or,

$0 = 0$.

That also works!

I'm a super genius!

Unfortunately, mathematicians had to further confuse the issue. They went on to claim that some infinities are bigger than others.

Intuitively, I would have accepted this.

Consider the *set* of positive *natural* numbers.

Let's exclude zero to simplify things.

1, 2, 3, 4, 5, 6, 7…

Now consider the *set* of *even, natural* numbers. Again, excluding zero.

2, 4, 6, 8, 10, 12…

There seems to be more total *natural* numbers than just *even* numbers, because the *set* of *evens* leaves out the *odd* numbers. Right?

Wrong!

Mathematicians claim they are equal infinities because they can be paired up one-to-one as you go out forever.

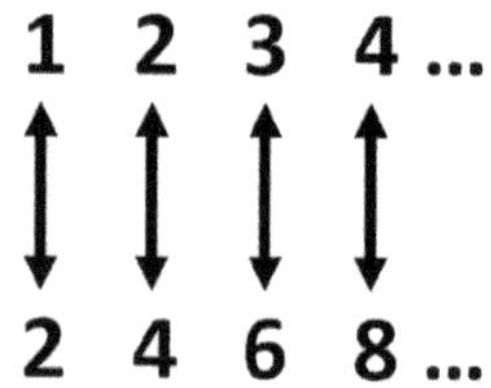

Nice trick.

Meanwhile, what about the *set* of *rational* numbers or fractions?

It's commonly accepted that there are an infinite number of fractions between two *natural* numbers. Afterall, no matter how infinitesimal the fraction, you can always find one smaller. It's similar to the proof that there is an infinite number of *natural* numbers.

Consider the smallest fraction you can imagine and call it $1/n$. You can always find one smaller, expressed as $1/(n+1)$.

As an example, compare 1/1,000,001 which is smaller than 1/1,000,000 simply because the *denominator* is bigger. Even though the difference is only 1. You can always create one more.

So, there's no way to pair them up. Right?

Sorry. I can't say.

Mathematical proofs that show some infinities to be bigger than others are extremely complicated and mathematicians have defined more symbols and rules for this. They talk about different *orders* of infinity and introduce concepts such as *bounded*, *unbounded* and *potential* infinities which is way beyond my brain power.

I would prefer to say that infinity is a *concept*, not a number, and not something that makes sense in arithmetic calculations.

An infinitely better solution.

Ø

Chapter 24: Symbols

Throughout this book, we've repeatedly made use of commonly recognized symbols such as $+ \div \sqrt{} - \pi$, etc.

In fact, at one time, such symbols were unknown. Instead, mathematicians described operations with words, which they also claimed prevented misunderstandings.

So, $(1 + 2) = 3$ might have been transcribed as, "Taking one unit and adding two units to it, results in three units."

Think how this might have sounded when solving an equation such as $(3a + 1) = 16$.

"How would you divide 16 apples equally among three people if one of them is rotten, so that each person gets a fresh one?"

Symbols are important in mathematics, as well as in the overall human experience. Not just as numerals or operators but also as concepts.

Most organizations, be they religious, political, corporate or otherwise use them. Think of the cross, the star and the crescent.

Companies call their symbols *logos*. Products call them *trademarks*. Countries call them *flags*.

They are just a shorthand way of summing up their beliefs or attributes. If you see *7Up* on the side of a can, you will most likely expect to find a soft drink inside that is clear and bubbly with a lemon-lime taste.

When you see "golden arches" on a store, you're likely to think of *McDonald's* and the various food products as well as the fast service experience inside.

Symbols communicate across languages and cultures. You typically know which public restroom to use around the world even if you don't speak the language. You know that a red light means you need to stop your car at an intersection.

Anywhere in the world.

Obviously, the symbols for numbers as well as their arithmetic operations have become a universal language.

Today's math symbols have both numerical or quantitative meaning as well as conceptual or abstract meaning.

The numerals we use today represent quantitative concepts that can apply to any object. So, we can say one apple, one rock, one dog etc. However, some ancient civilizations used different symbols to represent the *identity* of an object as well as its quantity.

So, two rocks might have been represented as O O, two humans as :) :) and three dogs as <^^ <^^ <^^.

If you haven't guessed, I made these up. And if you didn't recognize the "dog numeral" imagine the snout and ears.

When we perform arithmetic calculations or solve equations, we use numerals in the abstract. We don't need to assign them to anything if we don't want to.

We've already noted how symbols such as π can represent concepts or properties that can't be expressed as definite numbers within the traditional mathematical system.

We've shown how these symbols can be used arithmetically such as $\pi/\pi = 1$, even though we can't exactly express the value of π itself.

Similarly, $(\sqrt{2} - \sqrt{2}) = 0$ while $(\sqrt{2} \times \sqrt{2}) = 2$.

The symbols are shortcuts for combining elements of a *set* that make mathematical computations more efficient.

Take multiplication. As noted, multiplication is just a notational shortcut for addition.

You can write $(3 + 3 + 3 + 3)$ as (3×4).

However, it gets more complicated when you combine different operators.

Just think about how cumbersome it would be to write $(3 + 3 + 3 + 3) \div (2 + 2 + 2 + 2 + 2 + 2)$ vs the much familiar $(3 \times 4) \div (2 \times 6)$ or even simpler $(3 \times 4)/(2 \times 6)$.

It's even more complicated when solving for an equation. Even a simple one.

Take for example $(4a - 1) = 7$. Obviously, $a = 2$.

But how easily would you see that solution if it were written $(a + a + a + a) - 1 = 7$?

It's almost like going back to Roman numerals.

And think about how convenient the notational shortcut is for a number multiplied by itself several times, called an "exponential" in traditional math.

As we know, $(2 \times 2 \times 2 \times 2)$ can be represented as 2^4.

Of course, this asks why isn't there a notational shortcut for subtraction?

Why can't we shortcut $(2 - 2 - 2)$ as 2 followed by some symbol?

It would seem logical to represent that as 2^{-3}.

However, someone already *defined* 2^{-3} as $(\frac{1}{2} \times \frac{1}{2} \times \frac{1}{2})$ which equals $\frac{1}{8}$.

So, let's try using the symbol Ø as a shortcut to describe the operation $A \times (B + 1)$.

Here's an example.

Consider the equation where $A = 2$ and $B = 3$ and solve for a. So, $a + [A \times (B + 1)] = 12$.

This gives $a + [2 \times (3 + 1)] = 12$, which means $a = 4$.

Now consider the expression $a + [2 \text{ Ø } 3] = 12$. This translates to $(a + 8) = 12$, or $a = 4$.

Seems to work.

Let's try $a - [A \times (B + 1)] = 2$ where $A = 2$ and $B = 3$ again.

This gives $a - [2 \times (3 + 1)] = 2$, so $a = 10$.

Now consider, $a - (2 \text{ Ø } 3) = 2$. This gives $(a - 8) = 2$ which also results in $a = 10$.

Still works.

So, hopefully, the Ø symbol can lead to other breakthroughs within the traditional mathematical system like π or $\sqrt{2}$ which can't be written as definite numbers, as well as shortcut symbols like ÷ × ! that can be used arithmetically to more easily solve equations.

Of course, this raises the interesting question of why mathematicians didn't create a symbol for 0/0 or any number n, $n/0$ that can work within their system.

Very symbolic.

Implications: Natural Math

Ok. All this stuff may be interesting or even irritating to traditional mathematicians, but what's the point of *division by zero*?

Really, what *is* the point?

Besides its value as a notational shortcut, there are several points.

Nature.

Creativity.

Freedom.

Evolution.

Fundamentally, traditional mathematics as we know it claims to describe nature.

It has never pretended to tell us the "why" of the natural world, but has insisted that it is the ultimate descriptor of it.

Initially, it may have just started out as a tool for counting that helped us complete simple day-to-day transactions. But later it became the basis for all the sciences, be they natural, social or economic.

However, I often feel it tries to reverse-engineer nature by coming up with a mathematical construct that it likes, then desperately searches the natural world for proof that validates it.

The search for the *Golden Ratio* in nature is just one case in point.

Importantly, it is very critical to remember that the traditional mathematical system is based on definitions and *axioms* or "self-evident" truths. Subsequently, the system employs "if-then" logic to expand it.

Conveniently, if something doesn't fit, the system introduces new definitions, concepts and symbols such as $\sqrt{2}$ and i.

"If-then" logic is held up as the ultimate standard of intellectual thinking, but this assumption has its own limitations because it defines only two states of being.

This or That.

On or Off.

Black or White.

No shades of gray.

It also attaches a causality to actions operating on events that may not always exist.

If there were perfect mathematical causality between events, then weathermen would not need probability to produce a forecast.

Of course, they would say that the many variables involved in weather and all the interactions among them are beyond our current computing capabilities.

At the same time, this sort of "if-then" thinking defines only a small bit of man's experience. Also, while it can "prove" a lot of things, it says little about the truth or reality of them.

Consider the following.

Someone says to you, "I am lying. So, now tell me whether I am telling the truth or not."

If you assume they are lying, then you can conclude they are telling the truth.

If you assume they are telling the truth, then you can conclude they are lying.

However, if you don't assume anything, then you can't conclude whether they are lying or telling the truth.

Proof does not mean Truth.

Ask any lawyer.

Also, a causal view of world affairs clearly indicates that things don't always occur logically and man does not always behave rationally. Many great decisions are made using intuitive judgement rather than mathematical models driven by algorithms and big databases.

Even the cherished *quadratic* equation requires a judgement call to pick the most appropriate of two solutions when applied to a real-world problem. And not just when one solution is a *negative* number.

Some years ago, while a graduate business school student, I attended a class on quantitative analysis and decision-making.

The professor took us though a case study for a shipping company that wanted to know the most efficient way to manage its transport logistics. After analyzing the data and applying a mathematical model with this objective, the solution was very interesting.

It indicated that the most efficient solution was to periodically leave a ship stranded at sea for lack of fuel.

The professor smiled when he presented the solution.

The lesson was clear. There is a limit to what quantitative, big-data modeling can provide.

We are now living in an age of *Big Data*. The premise is that the more data you have, the more you are able to predict or even influence an outcome.

The upshot has been the quest for more data, more complex algorithms and more computing power.

Yet, I still can't rely on the weather forecast. And when I open my email homepage, I typically get hit with ads for women's cosmetics and feminine hygiene products.

I would have thought that my first name alone would have told their algorithm to send me beer ads.

Additionally, my online searches are overwhelmingly related to travel and culture.

So, shouldn't I be getting airline and hotel ads?

Maybe I'm missing *Big Data's* big picture.

At the same time, all these computations are still based on a mathematical system that can't attach a number to something so simple, natural and ever present as the mathematical relationship embodied in the ubiquitous circle. We still need to express it as a symbol.

Namely π.

It is also ironic that the *Golden Ratio*, which is the mathematician's definition of natural beauty, is an *irrational* number that can't be expressed as the *ratio* of two *natural* numbers.

In this regard, *ratios* seem to be especially important in defining concepts in nature. Take any geometric shape, be it a circle, rectangle, triangle or whatever. Increasing or decreasing its sides by the same *ratio* or proportion maintains the integrity of that shape.

In a world of change, *ratios* offer something constant. Is this a hint from nature? Is it possible to construct a number system using *ratios* as a *Base*?

So, a key contention of *Division by Zero* is that the current mathematical system we know and love, which has served us well for hundreds of years, may not totally reflect or be embodied in the natural world that we observe or experience.

If it were, why did it need to introduce such unnatural elements as *negative*, *imaginary* and *irrational* numbers to make it work?

Maybe our ancient mathematicians got the initial definitions, rules and relationships fundamentally wrong. So, maybe our modern mathematicians need to rethink definitions, revisit their axioms or generally accepted, "self-evident" unproven beliefs, and come up with an alternative system grounded in our natural world.

Hence my call for *Natural Math*. The idea is that math should behave according to nature, not that nature should behave according to an artificial system of mathematics devised by man.

Conceptually, *Natural Math* suggests that the world is based on whole numbers, specifically *natural* numbers, that interact within a system of relationships that we may have yet to fully discover or construct.

I think that science fundamentally accepts this.

Otherwise, why have scientists spent millions of dollars looking for the "fundamental particle" that makes up the universe? They are not looking for the "fundamental fraction" of an atom or an abstract concept like infinity, but a unique, discrete particle.

Doesn't this tell us that they really believe the natural world is best described by whole numbers?

Natural numbers.

The amazing strides in technology over the last decades would also say so.

We live in a digital world, which is based on computing systems that only use two numbers.

Two *natural* numbers.

Zero and one.

Of course, this computing system is based on binary "if-then" thinking. Only two states of an electrical circuit. On and off.

It has been a tremendous stepping-stone to scientific progress. But now, maybe *quantum* computing, which reportedly allows for other states within its system, will launch the next *quantum leap* in computation and scientific achievement.

Pun intended.

Obviously, creative thinkers need the freedom to explore ideas that completely question the fundamental principles of an existing system without being derided and shut out.

But freedom is not enough.

Creativity must be nurtured.

The key to coming up with innovative, creative ideas is not just giving someone the freedom to act creatively, but also the *encouragement* to question everything and the *opportunity* to bring new ideas to life and implement them in our everyday world.

This does not mean that every innovative idea is worthwhile. However, all should be given a chance.

All ideas are innocent until proven guilty.

In the corporate world, I sometimes found that my boss or client would not let me take an idea into a meeting out of fear that their boss or boss's boss wouldn't like it. However, I believe that even blatantly "incorrect" ideas should also be presented because they provide good perspective even if they get rejected.

And some just might work.

You don't know if you don't try.

It took Europeans hundreds of years to give up Roman numerals and incorporate Hindu-Arabic numerals into their society. You'd think they might have learned from the fall of the Roman Empire, that Rome didn't have all the answers.

As noted, it's easier to cling to the constants in our lives because they provide security and peace of mind, but in reality, the only constant in the world is change.

Nature shows very clearly that those who adapt to change are the ones who survive and prosper.

Evolve or die.

Bottom line, *Division by Zero* is about challenging entrenched beliefs and systems, always looking for something better, using nature as a guide and evolving our intellects.

You've heard the phrases:

"Damned if you do. Damned if you don't".

"Catch Twenty-Two".

Now, when someone rejects an idea you have because it doesn't fit into their personal beliefs, operational systems, or organizational rule book, simply say,

"Division by Zero".

Appendix A: Out-of-the-Box

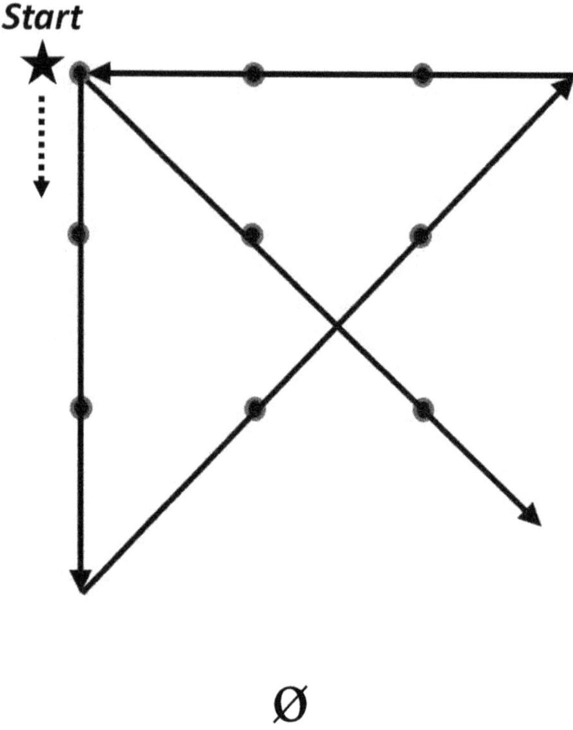

Ø

Appendix B: Golden Rectangles

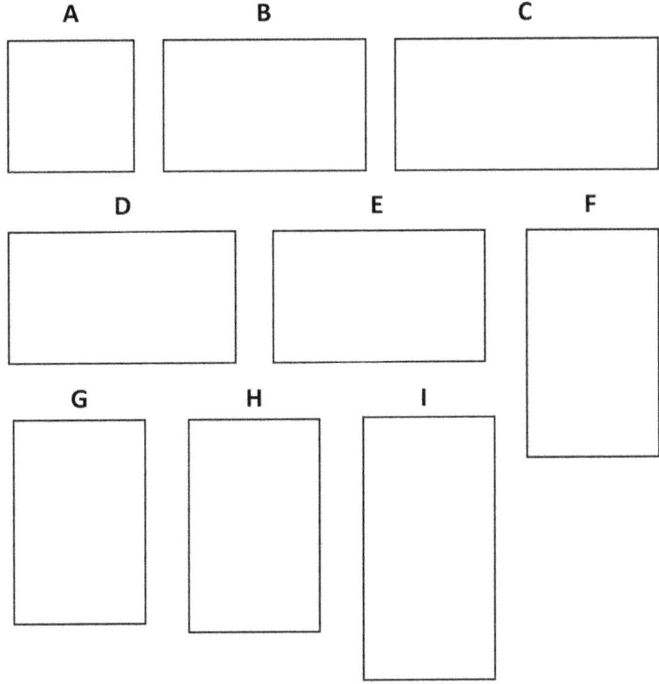

The *Golden Rectangles* are "E" and "H" which have the proportions of 1.6, which is roughly the value of $\varphi = 1.6$ rounded up to one decimal.

Appendix C: Number Operator Properties

Mathematicians would typically explore the relationship between a *set* of numbers and operations such as addition, subtraction etc. as well as *division by zero* in terms of the following properties:

- *Closure*
- *Identity Element*
- *Inverse Element*
- *Associative*
- *Commutative*
- *Distributive*

A collection of numbers combined by a defined operator that fulfills the first four of the above conditions is called a *group*.

Like *Set Theory*, there is a huge body of research that explores various *groups*, their properties and interactions called *Group Theory*.

Let's just get a little taste by investigating these conditions using *natural* numbers and see how *division by zero* makes out.

Closure means that combining any *elements* within a *set* using a selected operation produces a number that is also in the *set*.

Let's invent an operation symbolized by @.

Closure means that then if *A* and *B* are in the *set*, then so is *A* @ *B*.

So, *natural* numbers are *closed* under addition. An example for addition might be (3 + 2) = 5. Both 3 and 2 are in the *set* of *natural* numbers as is 5.

Note that *natural* numbers are not *closed* under subtraction. A very simple test is (2 – 3) = –1. But *negative* numbers are not in the set of *natural* numbers. They are in the *set* of *integers*.

However, the *natural* numbers remain *closed* under multiplication. So, $(3 \times 2) = 6$.

We can easily demonstrate that *natural* numbers are *closed* under *division by zero* since we already noted that $(A \text{ Ø } B) = A \times (B + 1)$. Therefore, as illustrated, *natural* numbers are *closed* under both addition and traditional multiplication.

Identity element means that for any number n, there is another number in the system, let's call it d, such that any number n, operated on by the *identity element* equals itself. That is $(n \ @ \ d) = n$.

In the case of *natural* numbers, the *identity element* is identified as 0 with respect to addition.

That is, for any number n, $(n + 0) = n$. For example, $(3 + 0) = 3$.

Similarly, for *natural* numbers the *identity element* is obviously 1 with respect to multiplication.

That is, for any number n, $(n \times 1) = n$. For example, $(3 \times 1) = 3$.

In the case of *division by zero*, the *identity element* is unsurprisingly 0. That is, for any number n, $(n \text{ Ø } 0) = n$.

For example, $(3 \text{ Ø } 0) = 3$.

Inverse element means that for any number n, there is another number or *element* in the collection, let's call it n^ such that n operated on by the *inverse element* equals the *identity element*. That is, $(n \ @ \ n\text{^}) = d$.

Careful not to confuse *inverse element* with *identity element*.

Note that the *natural* numbers have no *inverse element* under addition. Addition has an *identity element* of 0, so the *inverse* element for any *natural* number n, would need to be $-n$. That is, $n + (-n) = 0$. However, the *set* of *natural* numbers does not include *negative* numbers.

Traditional multiplication also doesn't work.

Remember that the *identity element* for traditional multiplication for any *natural* number n, is 1.

That is $(n \times 1) = n$ just like $(5 \times 1) = 5$.

The obvious *inverse element* for any *natural* number n, is $1/n$.

So, $(n \times 1/n) = 1$ just like $(5 \times 1/5) = 1$.

Whoops!

Recall that $1/n$ is a not a *natural* number. It is a *rational* number.

Meanwhile, *division by zero* has no *inverse element.* A quick look at the summary table in *Appendix D* shows there is no number such that for every n, $(n\ Ø\ n^\wedge) = 0$.

Associative means that pairing numbers together in a series of the identical operation delivers the same result irrespective of how they are paired and operated on.

So, $A + (B + C) = (A + B) + C$.

As you'd expect, addition and multiplication are *associative,* as in, $3 + (1 + 2) = 3 + (3) = 6 = (3 + 1) + 2$.

Meanwhile, $A \times (B \times C) = (A \times B) \times C$.

Therefore, $3 \times (2 \times 1) = 3 \times (2) = 6 = (3 \times 2) \times 1$.

However, *division by zero* is not.

A simple test indicates that 1 Ø (2 Ø 3) = 1 Ø (8) = 9 whereas (1 Ø 2) Ø 3 = (3) Ø 3 = 12.

So, the set of *natural* numbers under *division by zero* would not be considered a *group*, for what it's worth.

Now consider the remaining two properties which are not required conditions for a *group* though we use them very frequently in arithmetic.

Commutative means that the order of the numbers operated on doesn't matter. When I first heard about the concept of a *group*, I was surprised that this is not a required condition, because it was the very first number operator property I learned.

As you might expect, addition and multiplication are *commutative.*

So, $(A + B) = (B + A)$. And, $(3 + 4) = (4 + 3) = 7$.

Similarly, $(A \times B) = (B \times A)$. So, $(3 \times 2) = (2 \times 3) = 6$.

Meanwhile, subtraction is not. Note, $(A - B) \neq (B - A)$. Obviously, $(5 - 2) = 3$, but $(3 - 5) = -2$.

At the same time, some processes in nature or the real world are not *commutative*.

Putting on your socks first then your shoes, is not the same as putting on your shoes first then your socks. Commuters on your train to work will likely notice.

Traditional division is also *not commutative*. Hence, $(A \div B) \neq (B \div A)$. For example, $8/4 = 2$ whereas $4/8 = ½$.

And neither is *division by zero*. A simple test shows that (1 Ø 2) = 3, whereas (2 Ø 1) = 4.

Distributive means you can "distribute" the initial operator across numbers already grouped under a subsequent operation and get the same result. This is mostly associated with multiplication combined with addition or subtraction. We'd say that "multiplication is *distributive* with respect to addition and subtraction."

So, $A \times (B + C) = (A \times B) + (A \times C)$.

That is, $2 \times (3 + 4) = (2 \times 3) + (2 \times 4) = (6) + (8) = 14$.

And, $2 \times (4 - 3) = (2 \times 4) - (2 \times 3) = 2$.

On the contrary, addition is not *distributive* with respect to multiplication, $A + (B \times C) \neq (A+B) \times (A + C)$.

For example, $2 + (3 \times 4) = 14$, but $(2 + 3) \times (2 + 4) = (5) \times (6) = 30$.

Subtraction and division also fail inspection. A quick test for whether subtraction is *distributive* with regards to multiplication yields, $8 - (2 \times 3) = 2$.

However, $(8 - 2) \times (8 - 3) = (6) \times (5) = 30$.

Testing for whether division is *distributive* with regards to addition shows, $12 \div (2 + 4) = 2$.

However, $(12 \div 2) + (12 \div 4) = (6) + (3) = 9$.

Division by zero is not *distributive* with regards to addition.

A quick test finds that 2 Ø (3 + 1) = 10, whereas

(2 Ø 3) + (2 Ø 1) = (8) + (4) = 12.

Does this diminish its value as an operation?

Absolutely not.

I don't think that anyone is going to throw out addition or subtraction because they are not *distributive* with respect to multiplication or division.

Appendix D: Division by Zero Table

Ø	0	1	2	3	4	5	6	7	8	9
0	0	0	0	0	0	0	0	0	0	0
1	1	2	3	4	5	6	7	8	9	10
2	2	4	6	8	10	12	14	16	18	20
3	3	6	9	12	15	18	21	24	27	30
4	4	8	12	16	20	24	28	32	36	40
5	5	10	15	20	25	30	35	40	45	50
6	6	12	18	24	30	36	42	48	54	60
7	7	14	21	28	35	42	49	56	63	70
8	8	16	24	32	40	48	56	64	72	80
9	9	18	27	36	45	54	63	72	81	90

www.ingramcontent.com/pod-product-compliance
Ingram Content Group UK Ltd.
Pitfield, Milton Keynes, MK11 3LW, UK
UKHW021648190726
13853UKWH00001B/128

9 789887 515807